URBAN SKETCHES

ARNE SÆLEN

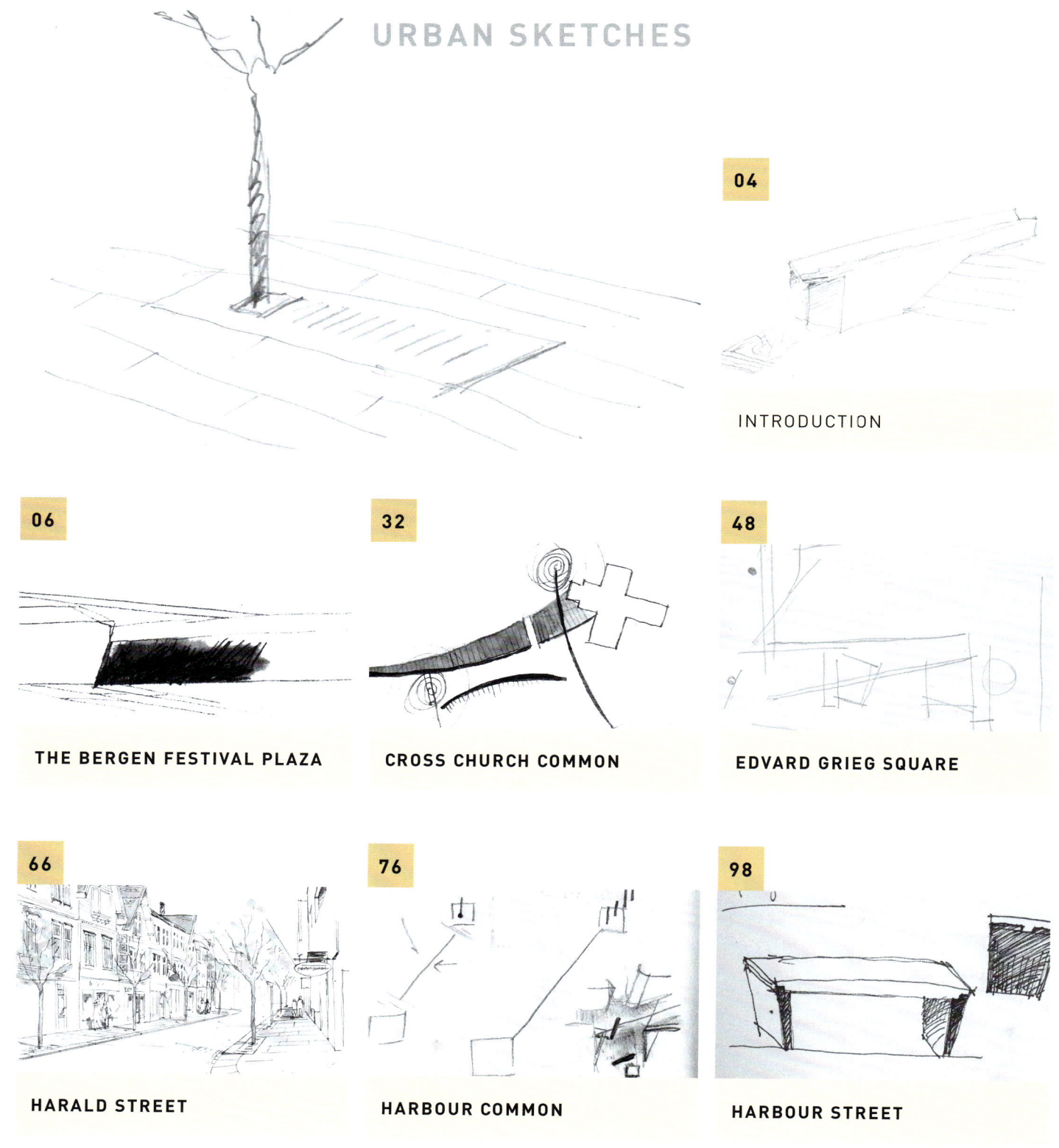

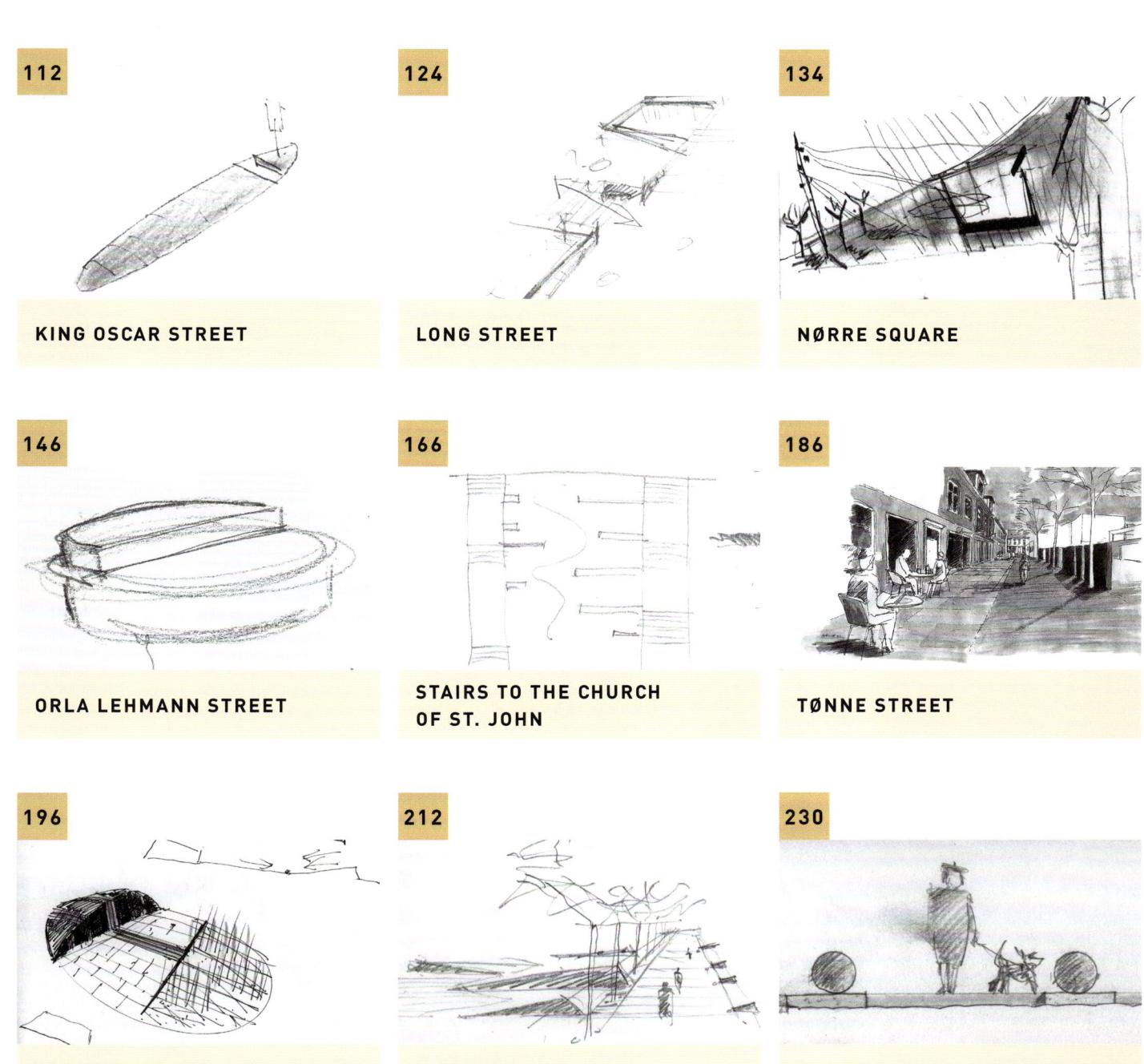

112 KING OSCAR STREET	124 LONG STREET	134 NØRRE SQUARE
146 ORLA LEHMANN STREET	166 STAIRS TO THE CHURCH OF ST. JOHN	186 TØNNE STREET
196 VEJLE THEATRE SQUARE	212 VINCENT LUNGE SQUARE	230 CREDITS & SPECIAL THANKS

How do one visualise an idea? How du you transform an idea into a physical state?

My way has always been sketching. A sketch may be a pencil drawing, as well as making clay or wooden models, or working on computers. But most of all I prefer using pencils on paper.

With a few lines in just seconds you can make mental diagrams that may convince others of your intentions. And later on these lines can develop, and in the end come into being as physical objects or sites. Mental diagrams will in this way, in the end, manifests into spaces where people can live their life.

Landscape sketches varies from the details of joints to large scale landscapes. This can also be done by computer models. But the sketch is still a much faster and colourful way of presenting ideas. 3D sketches is a much slower, but then again a more presise way of visualising. Using a lot of time in renderings, a computer model can almost look like the final physical project.

The pencil sketch can never do that. But the pencil sketch emphasises feelings and fantasies about how the final project may be. This is fundamental different from presenting a final picture.

Bergen, March 8

Arne Sælen

THE BERGEN FESTIVAL PLAZA

BERGEN, NORWAY

THE BERGEN FESTIVAL PLAZA is located along the city´s main axis, with majestic mountains in the distance. The Festival Plaza is situated close to a small lake, facing south. Massive, polished granite walls are bordering the lake. They resemble the mountains around the city the way they are tilted and separated. The edges are reinforced with stainless steel edges.

The plaza is slightly tilted against the lake to the south and towards the light reflections from the lake. The paving is designed in grey granite with veins. Because the plaza is only slightly tilted, we had to use open drains. Two drains are sunken 5 mm and they are constructed from massive stone, in which the polished drains are carved out. To remove the water, there are inserted four storm drains in each drain. The storm drains are custom designed and cast in bronze.

On both sides of the plaza, four gigantic steps in white granite lead down to the floor. The steps are connecting the plaza with the adjacent parks, and nine cherry trees are striding down the steps and into the plaza. Identical to the 1oo year old Kazan trees situated around the lake. The steps have different surfaces; every second step is rough picked, while the others have a natural surface. The outer 30cm of each step is made with a different surface to aid poorly sighted and blind people.

Sketch of the open drain.

3D sketch by MIR Visuals

The steps are made up by 101 elements, the largest weighing approximately 12 metric tons.

The Plaza ends in a 50m broad staircase leading down to the lake. Thus the interlocking water and land. By descending the stairs, one is transmitted into a completely different situation; leaving the pulsating city life and turning to the calm of the water.

The main lighting system is eight large poles, four on each side.

Between the granite steps, LED cables are inserted behind polycarbonate sticks. The steps are thereby literally floating in the night.

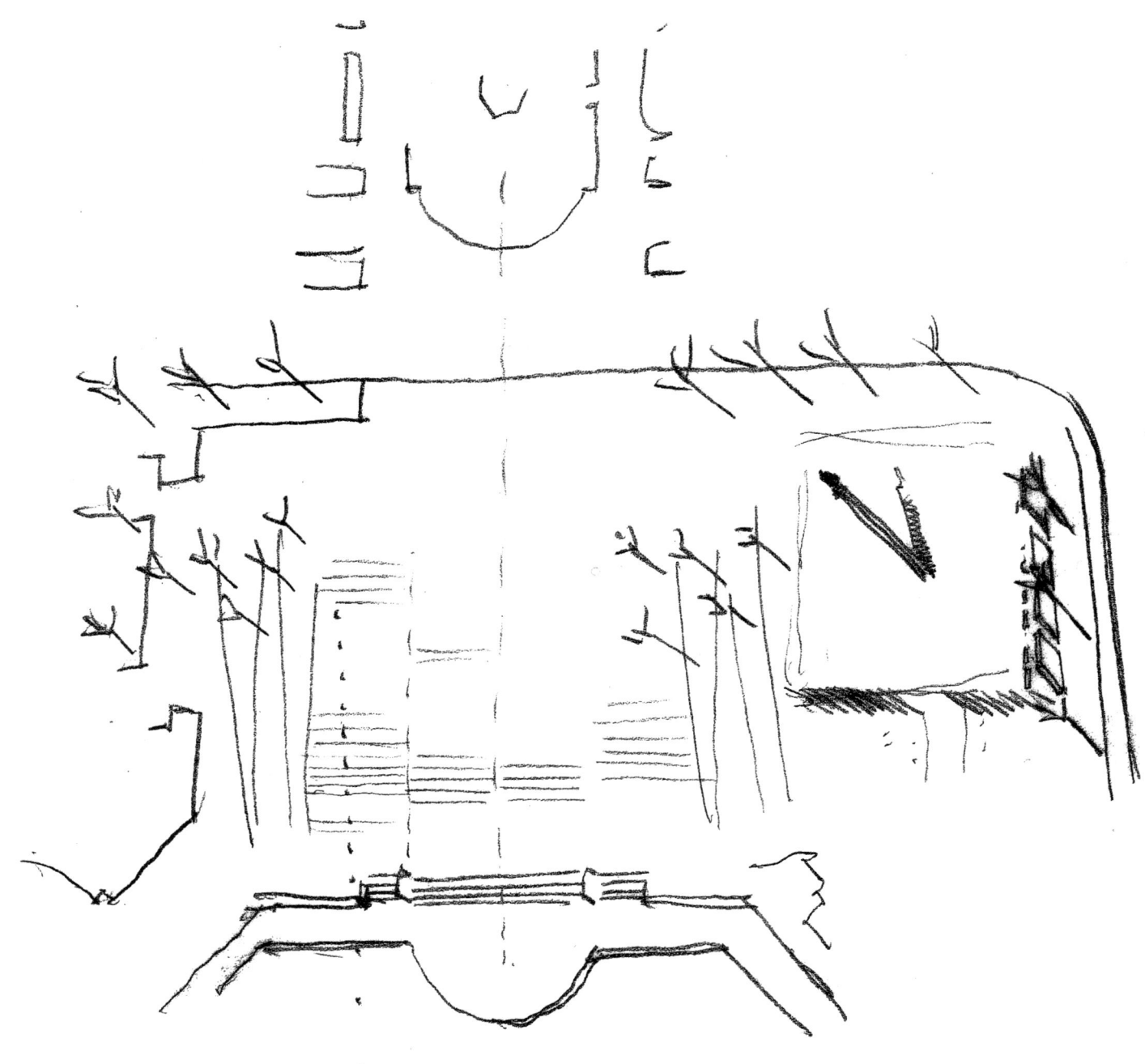

Early sketch with a trottoire close to the water.

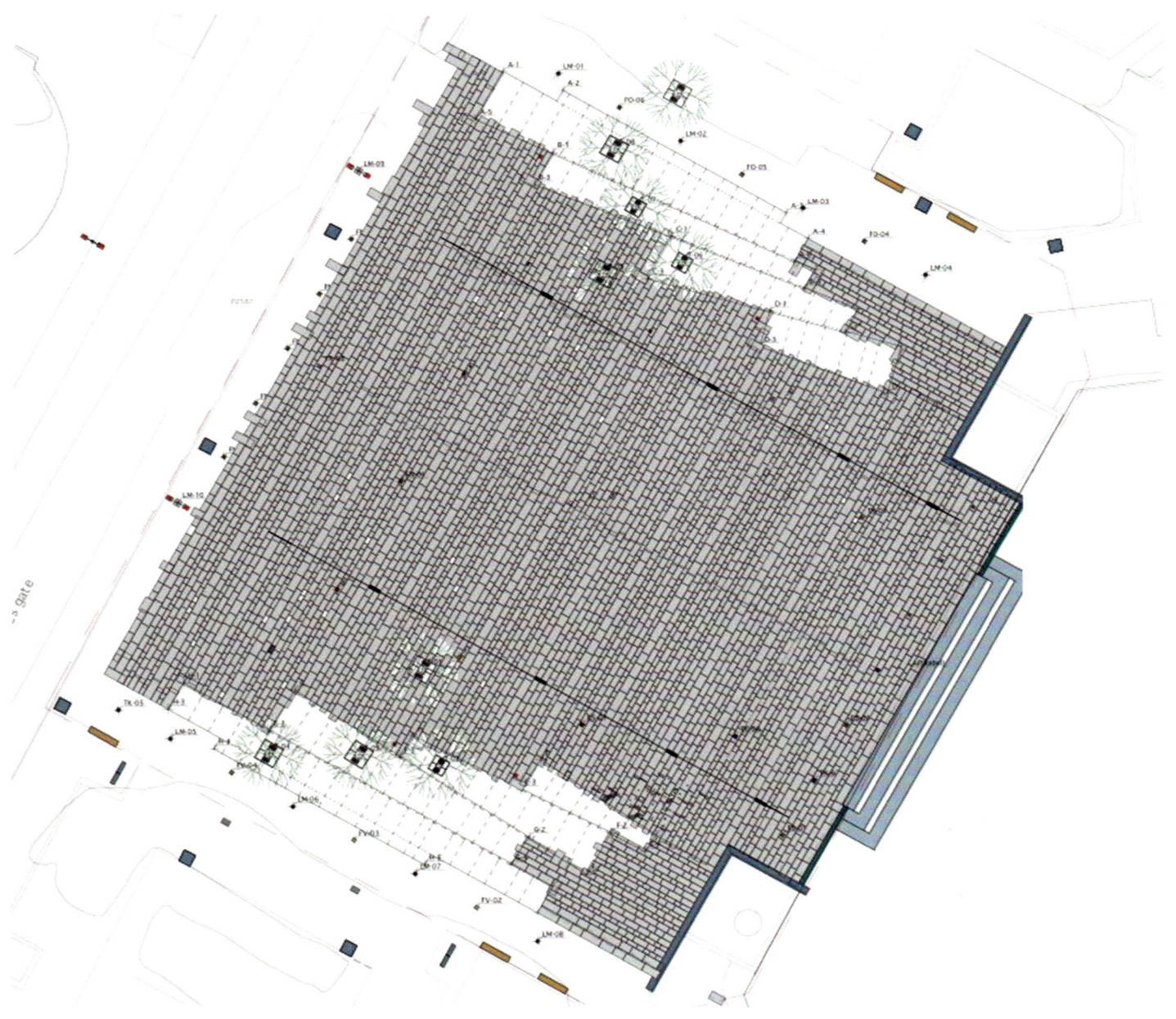

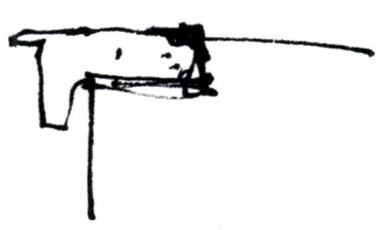

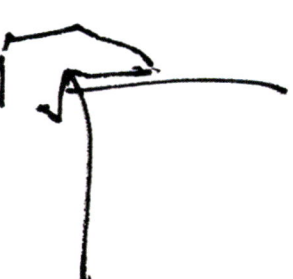

The Plaza ends in a 50 m broad staircase leading down to the lake. Thus the interlocking water and land. By descending the stairs, one is transmitted into a completely different situation; leaving the pulsating city life and turning to the calm of the water.

The main lighting system is eight large poles, four on each side.

Between the granite steps, LED cables are inserted behind polycarbonate sticks. The steps are thereby literally floating in the night.

21

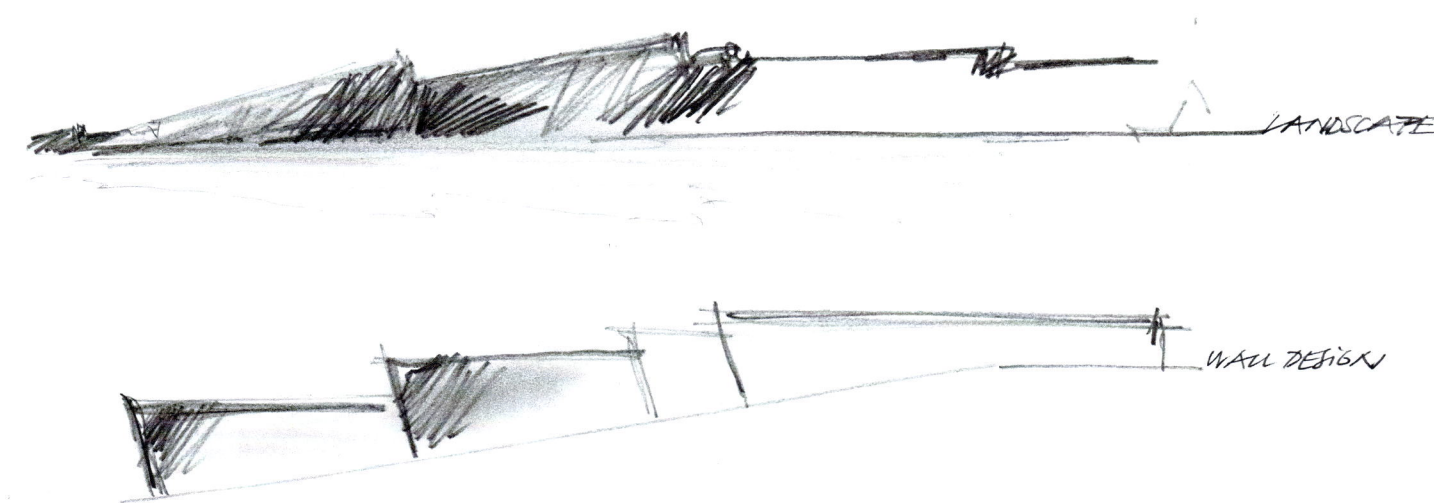

LANDSCAPE

WALL DESIGN

65 VP.
60 SP.

EDGE GRANITE STEP
EDGE TREE GRILLE
EDGE TREE GRILLE
EDGE GRANITE STEP

BIG GRANITE STEPS
CONCRETE LID FOR PLANTER
CONCRETE PLANTER FOR TREE

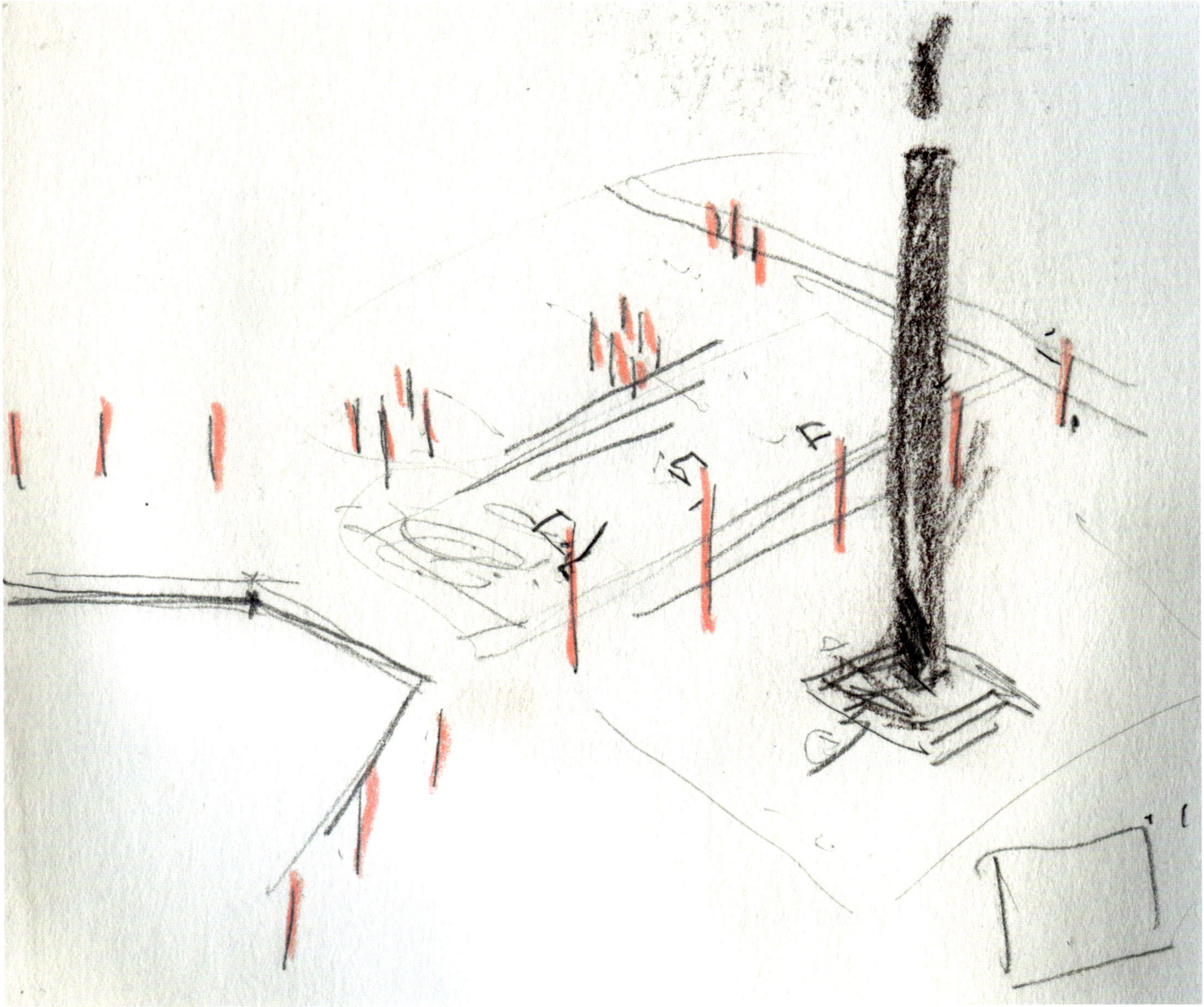

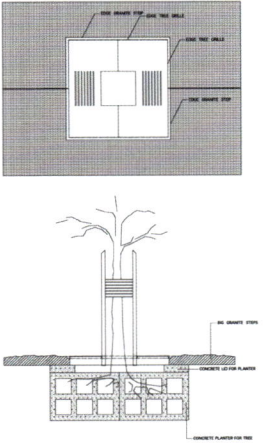

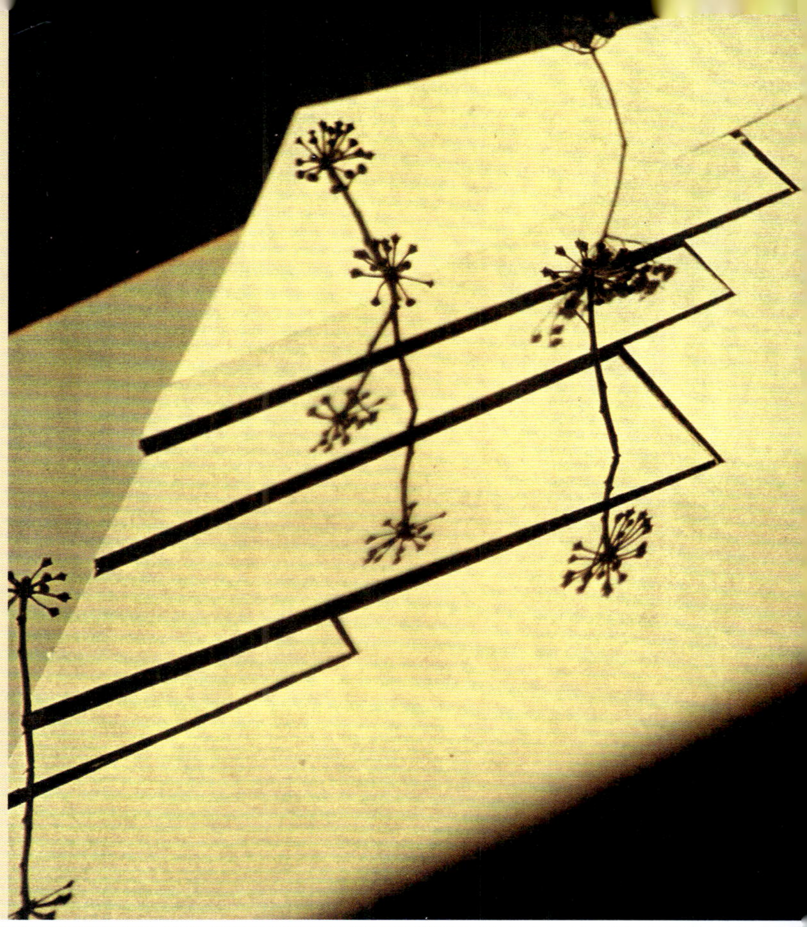

CROSS CHURCH COMMON

BERGEN, NORWAY

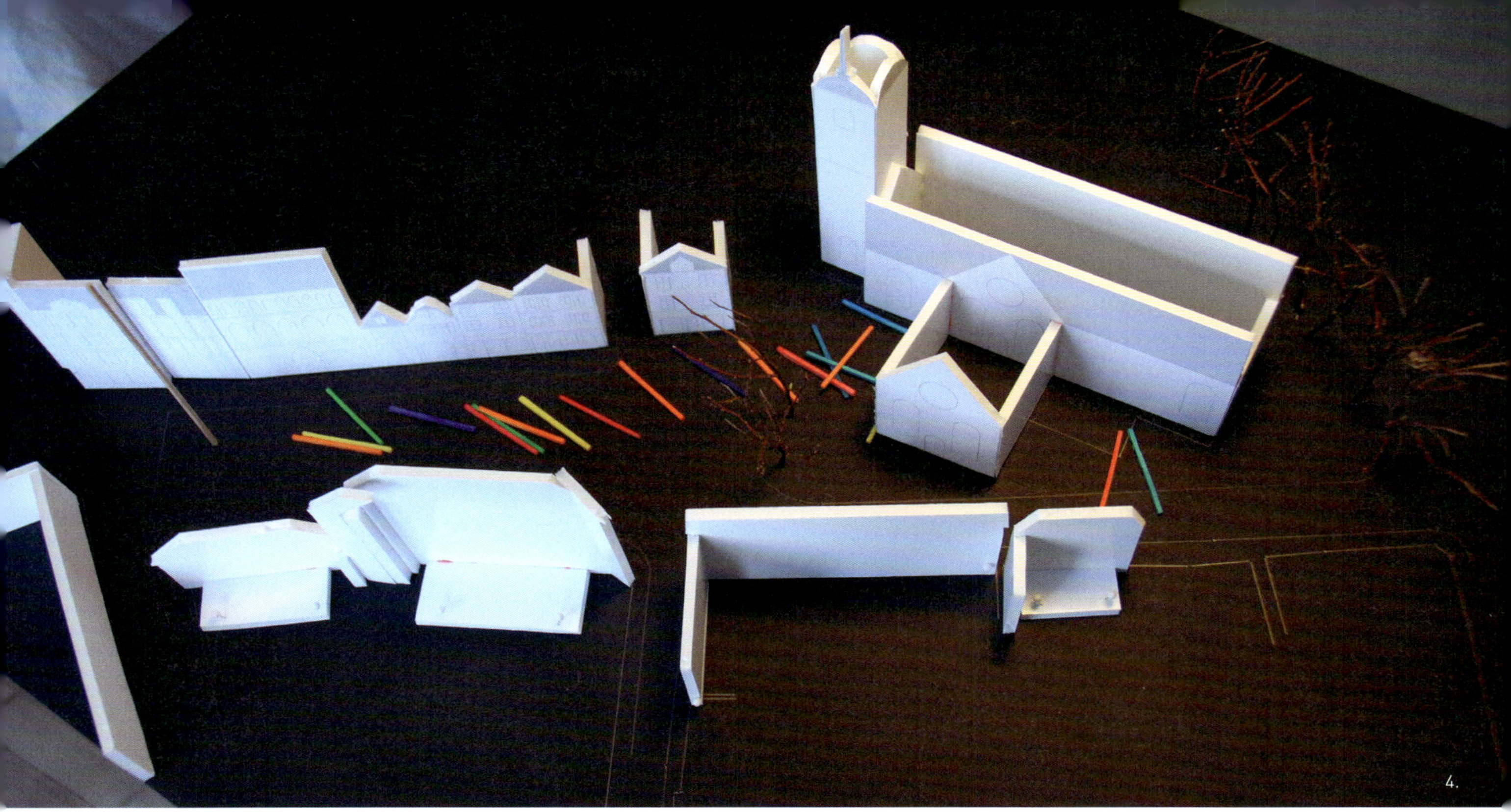

4.

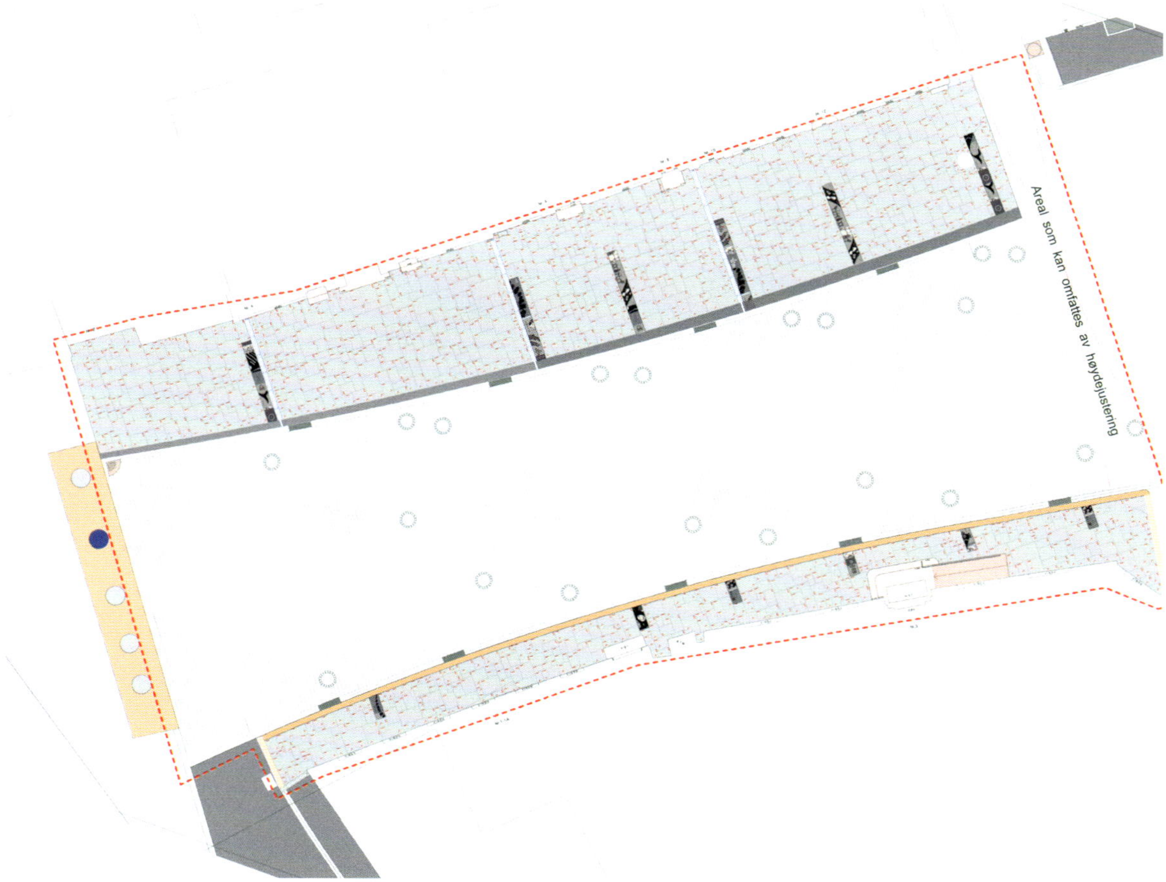
Areal som kan omfattes av høydejustering

2.

3.

1. Conceptual sketch showing the curved and extended pavement.

2. Preliminary sketch of the pavement intarsia.

3. Final intarsia.

4. Sketch model of the purposed dynamic movement across the common.

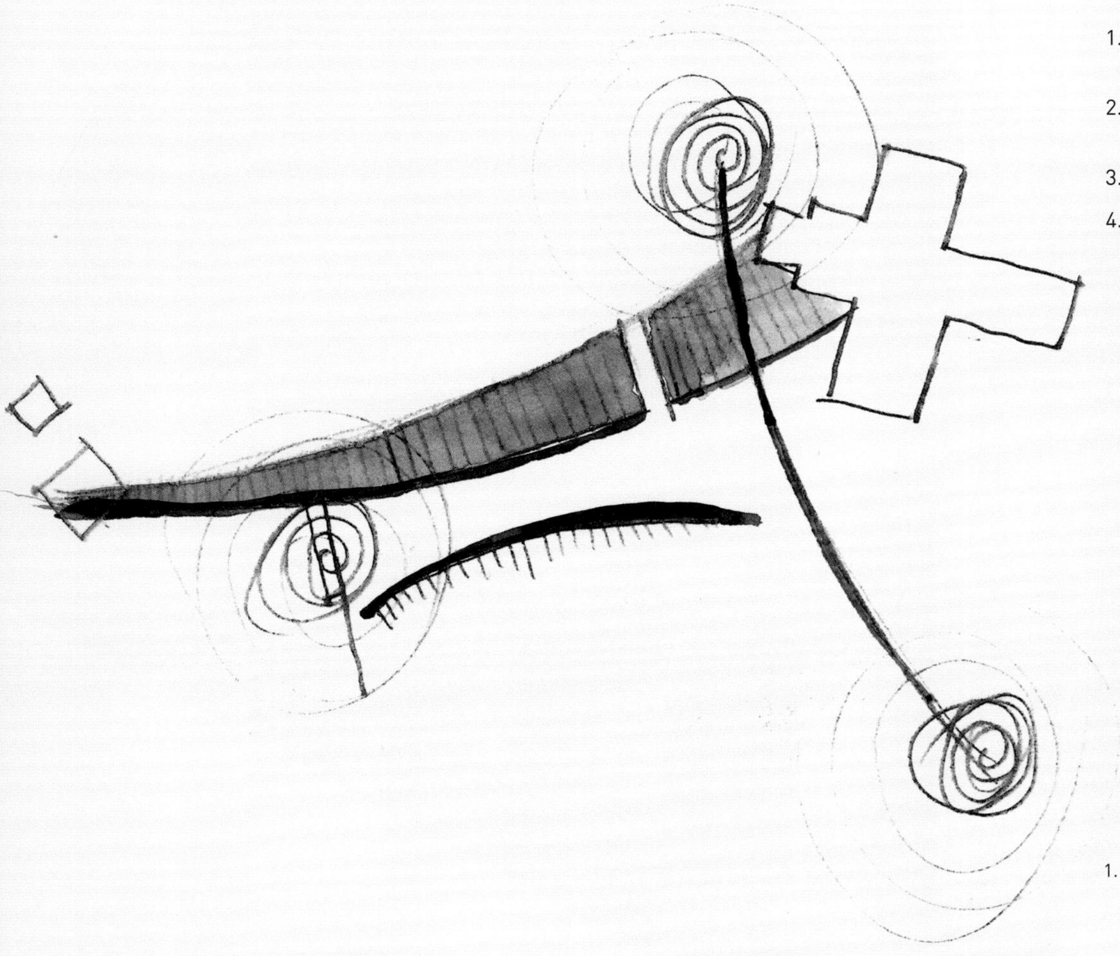

1.

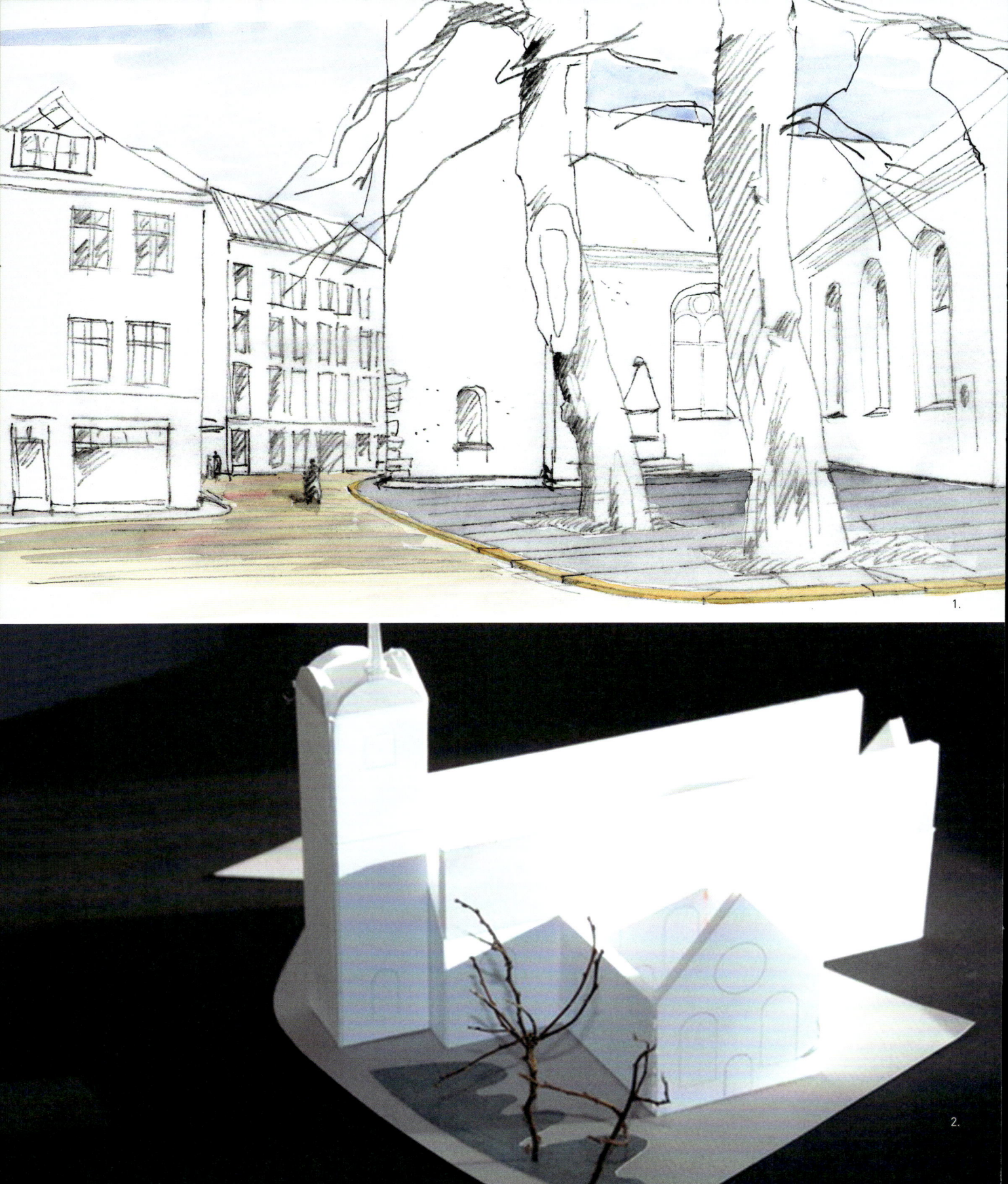

1.

2.

CROSS CHURCH COMMON is one of the oldest urban spaces in the city. It was the building site for the church in the 1150s, and became later one of the first commons in the city. The Cross Church Common, together with some adjacent streets, forms the heart of the medieval centre in Bergen.

The Church of the Holy Cross is dominating the square, more than 500 years older than the rest of the adjacent buildings. It therefore demands utterly respect.

All together the site is more than 4000 m^2. The increased and heavier traffic in the decades after WW2 have damaged the old paving, the oldest dating back to the 1750s. Thus all cars are banned from the area after the reinvention.

The main design was to open the squares lines by extending the pavings towards the church in a bold curve.

The new design shows pavement extensions, covered by slates. The pavement to the north, facing the sun got the largest extension. The former asphalt had to be removed on the pavements as well as from parts of the street. The cobbled stones were taken away for storage and were laid out in the same area as before.

As cultural signals and historical reminders, a new graphic design using medieval pottery, white marble, black granite and brass, were laid as intarsia in the pavements. These decorations are developed by the artists Kari Aasen and Lasse Berntzen, using fragments from both ancient pottery and modern water-jet cut granite.

The inspiration is a 3oo years old novel written by the author and poet Ludvig Holberg.

To prevent cars from driving through the common, a large granite form was established in the western end of the Common.

On top of this five gigantic clinkers was mounted; four in polished steel, and one i blue ceramics, created by Kari Aasen.

1. Conceptual sketch by the Church of the Holy Cross.
2. Alternative model of the same area.
3. Final extended pavement.

1. Proposal in front of the church. Not realised
2. Sketch of adjacent street. Not realised
3. Early sketch of the northern side of the common.

3.

Sketches from the adjacent streets.
Not realised

1.

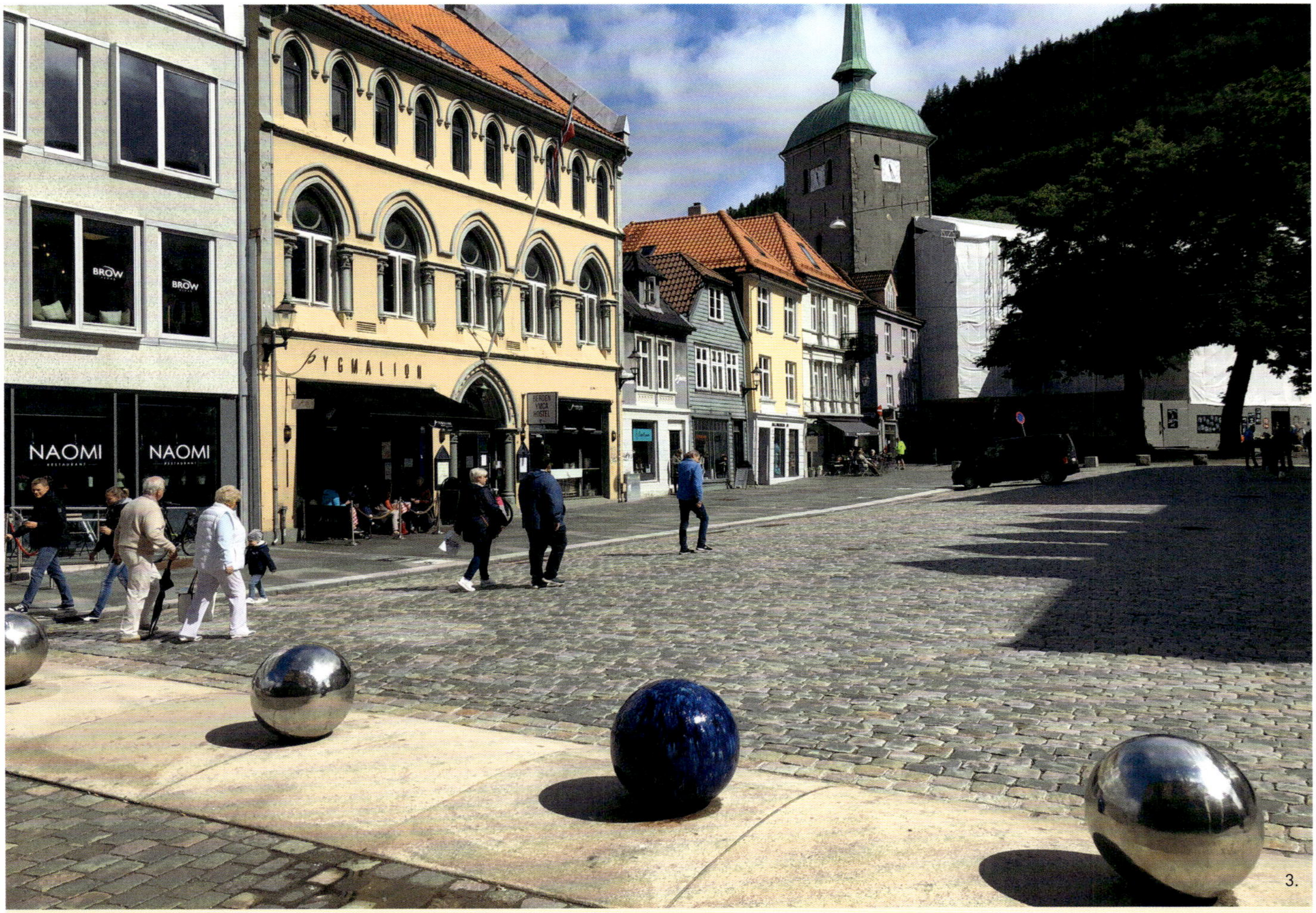

3.

1. Final sketch.

2. Preliminary sketch /section.

3. After intervention.

2.

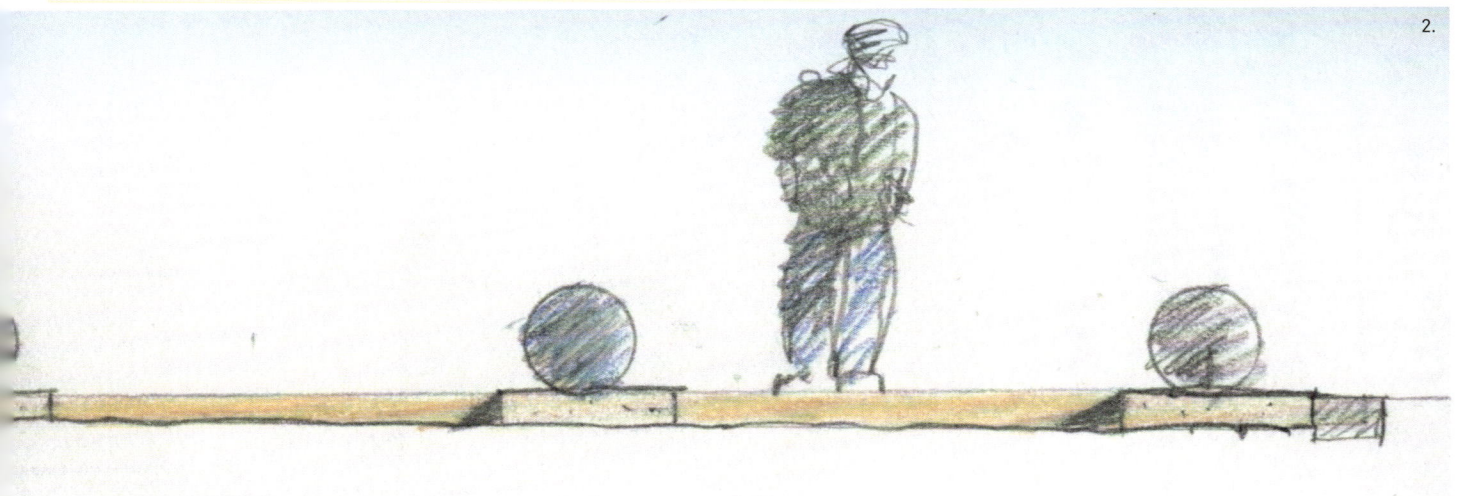

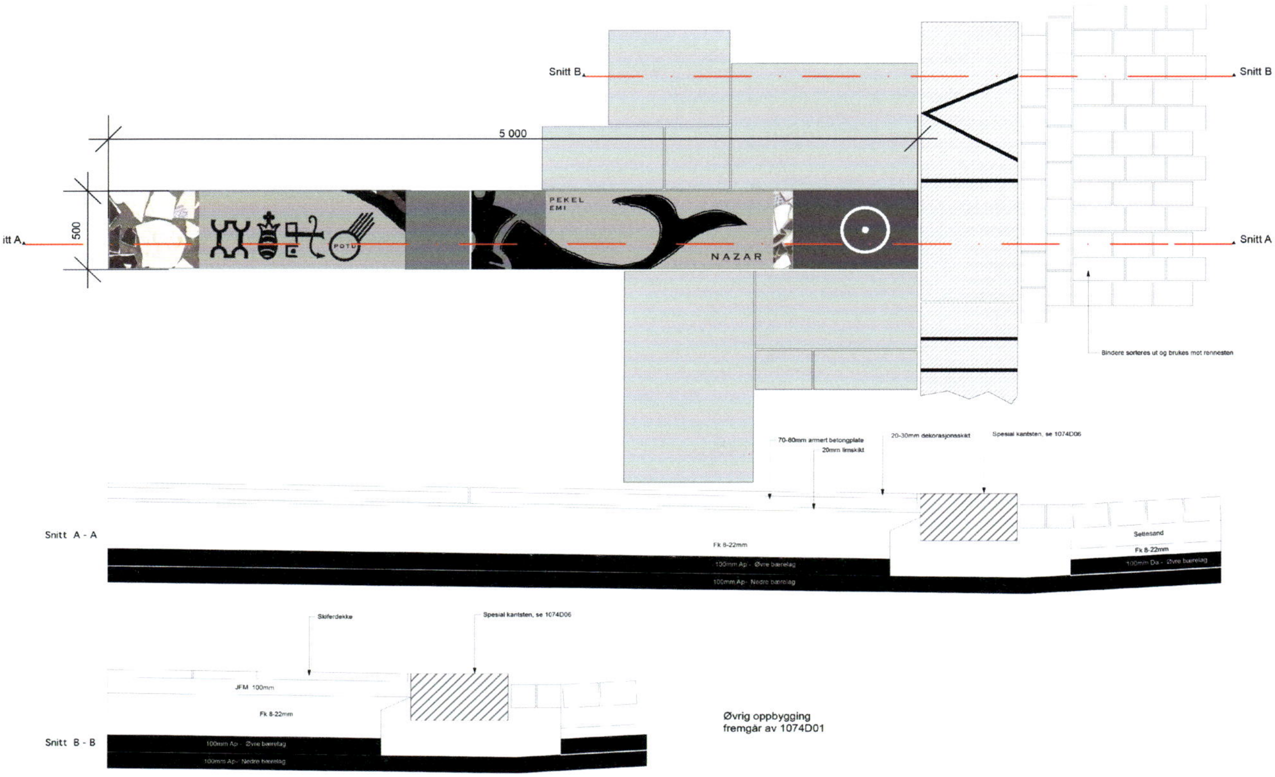

Snitt B
Snitt B
5 000
500
itt A
Snitt A
PEKEL
EMI
NAZAR
Bindere sorteres ut og brukes mot rennesten
70-80mm armert betongplate
20mm limskikt
20-30mm dekorasjonsskikt
Spesial kantsten, se 1074D06
Snitt A - A
Fk 8-22mm
100mm Ap - Øvre bærelag
100mm Ap- Nedre bærelag
Settesand
Fk 8-22mm
100mm Da - Øvre bærelag
Skiferdekke
Spesial kantsten, se 1074D06
JFM 100mm
Fk 8-22mm
Snitt B - B
100mm Ap - Øvre bærelag
100mm Ap- Nedre bærelag
Øvrig oppbygging
fremgår av 1074D01

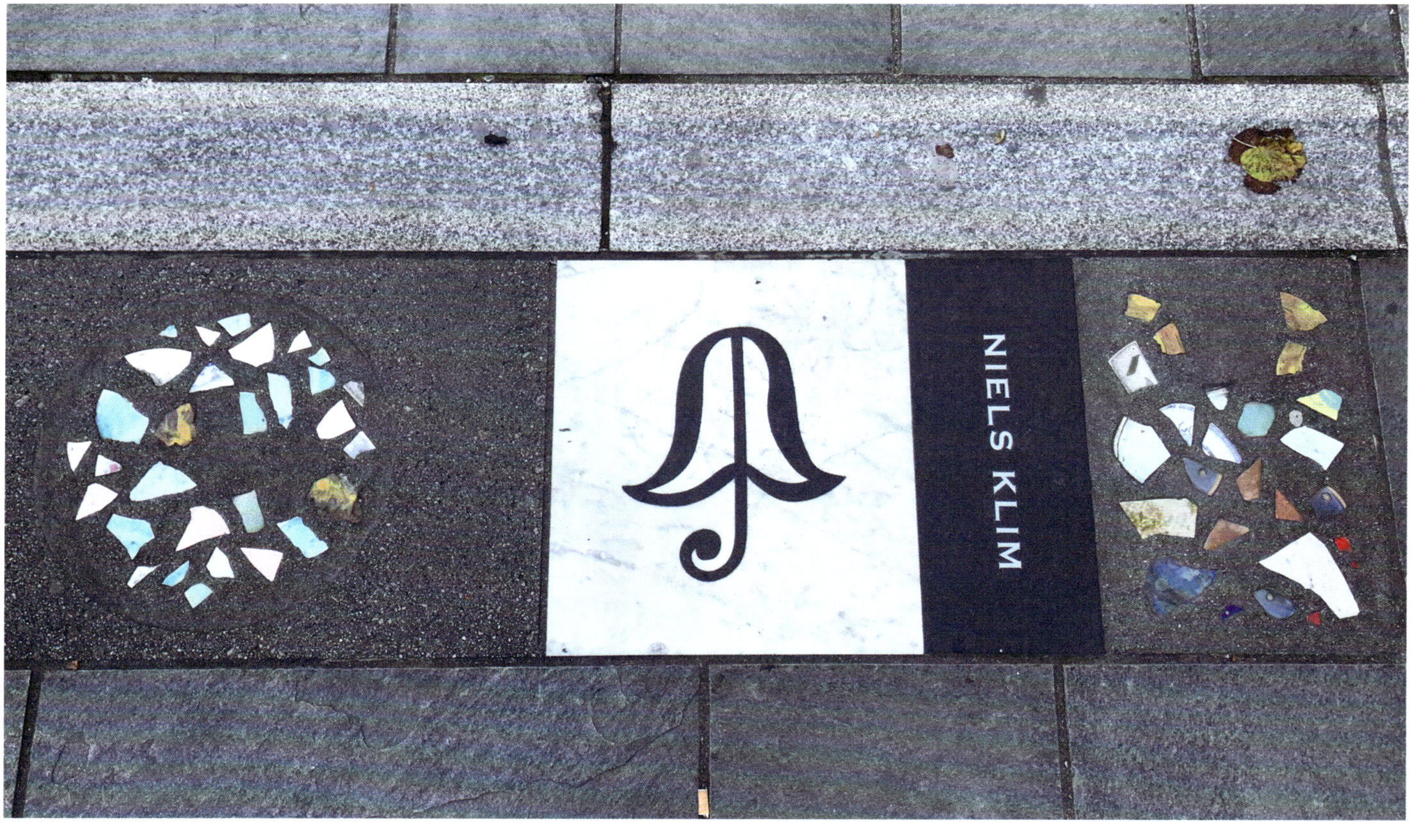

NIELS KLIM

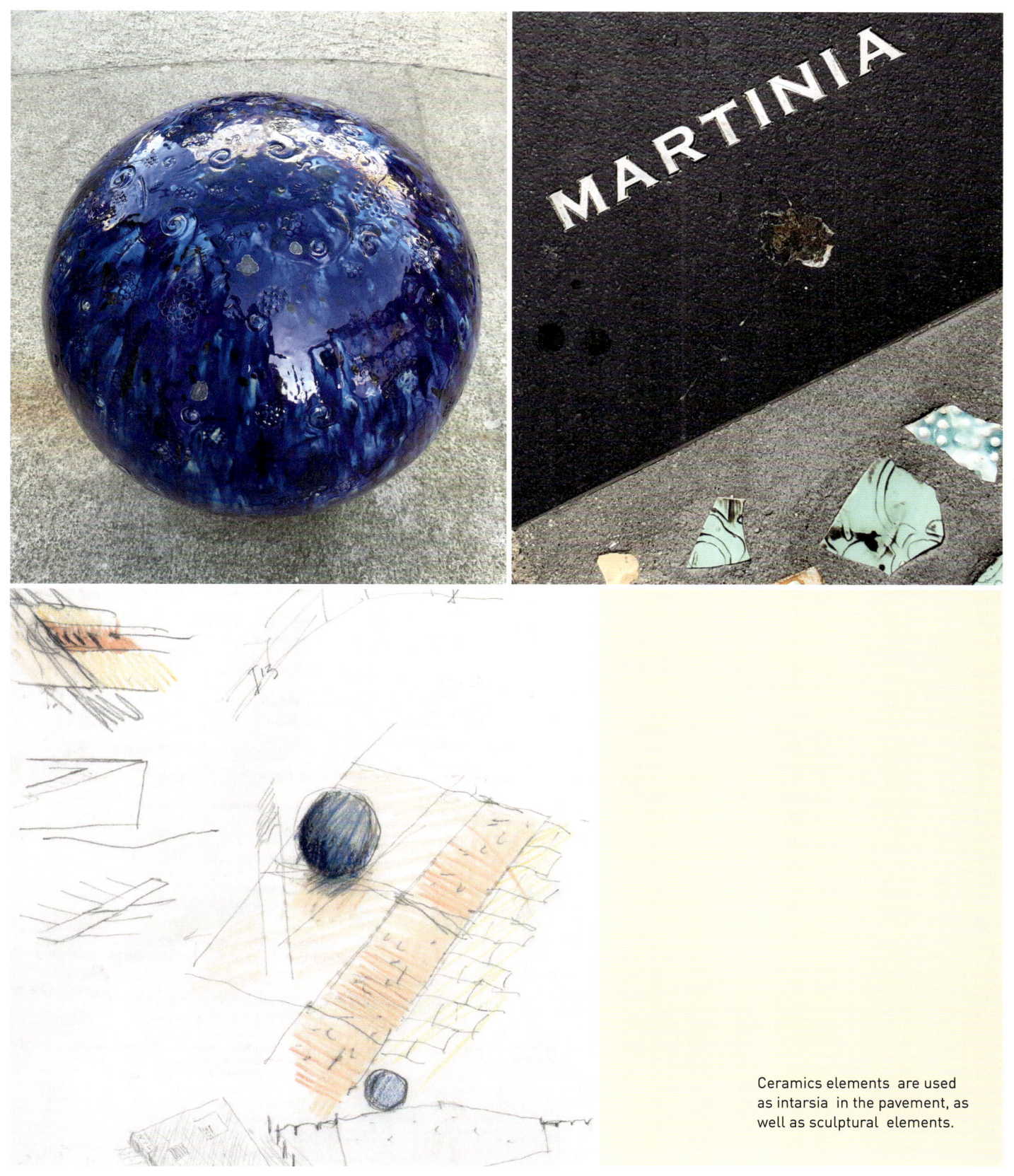

Ceramics elements are used as intarsia in the pavement, as well as sculptural elements.

EDVARD GRIEG SQUARE

BERGEN, NORWAY

Conceptual sketch

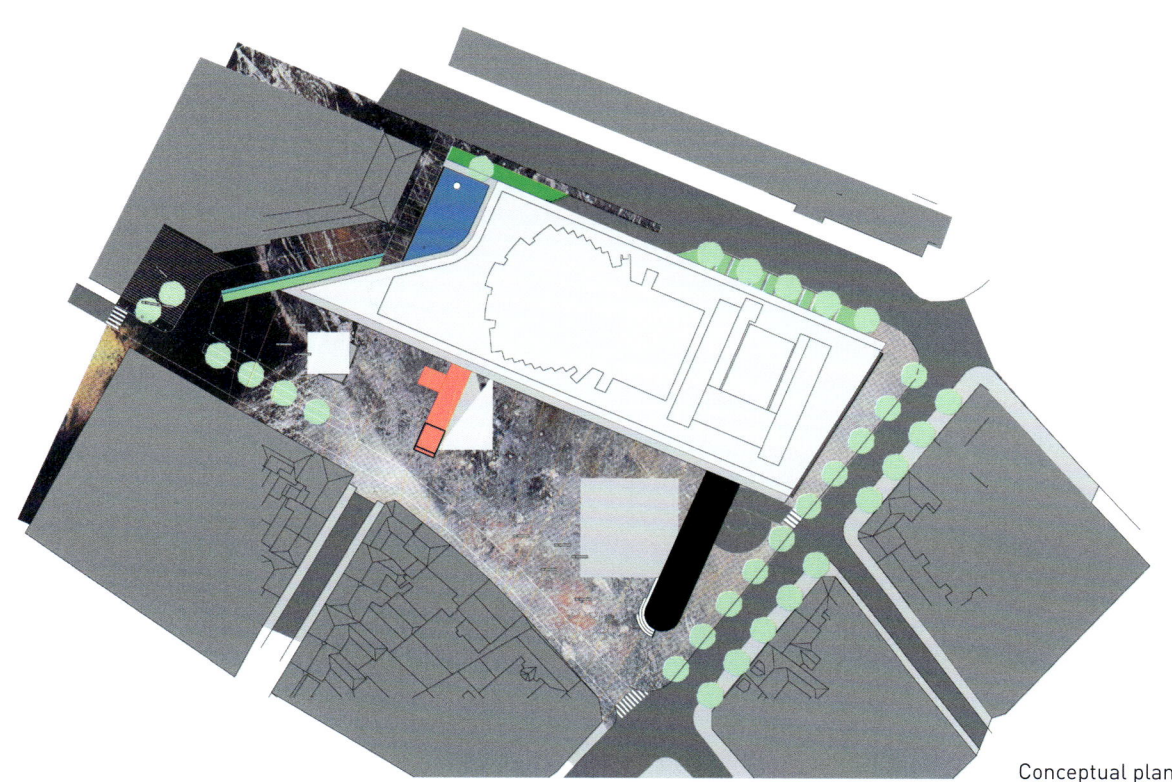

Conceptual plan

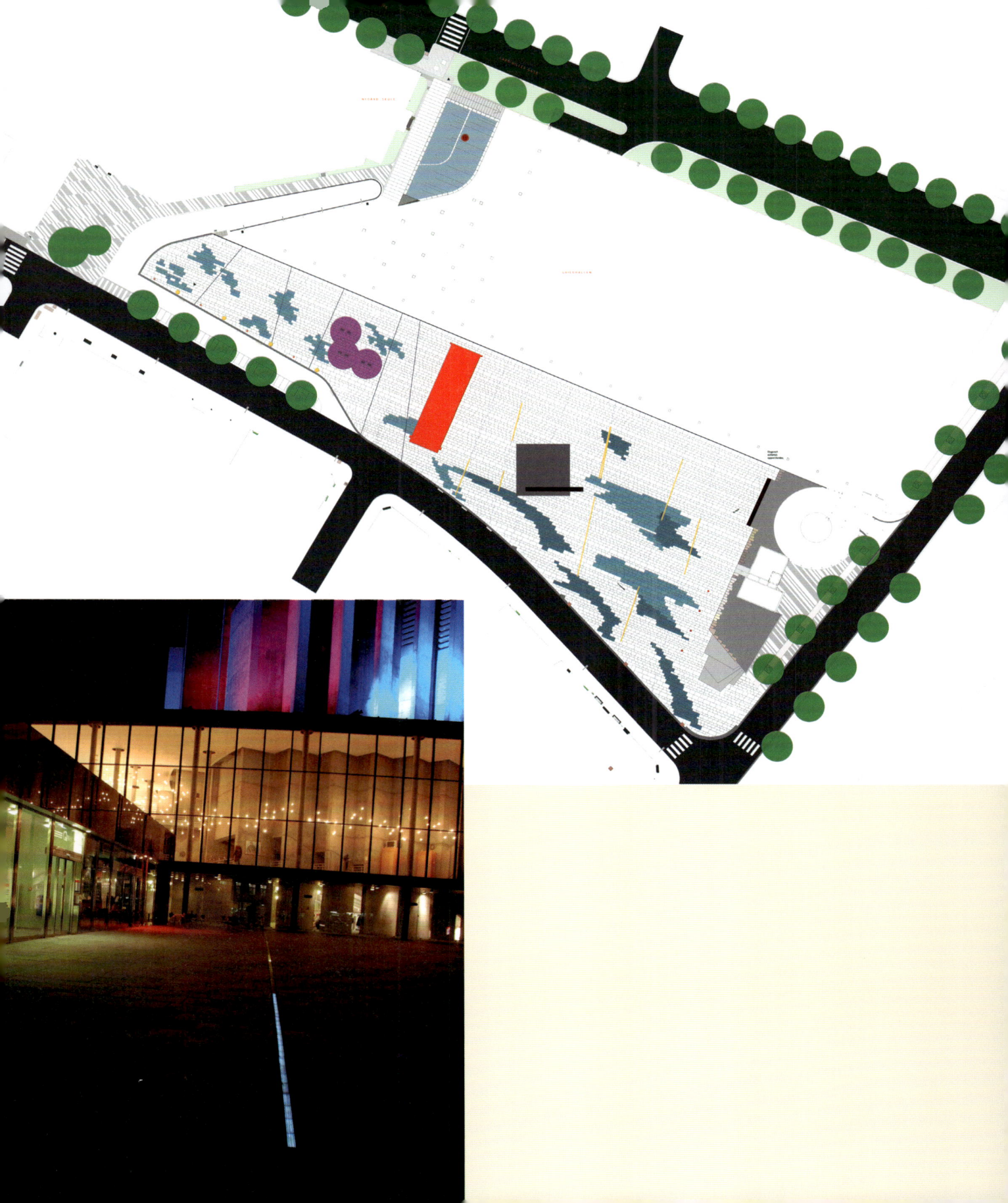

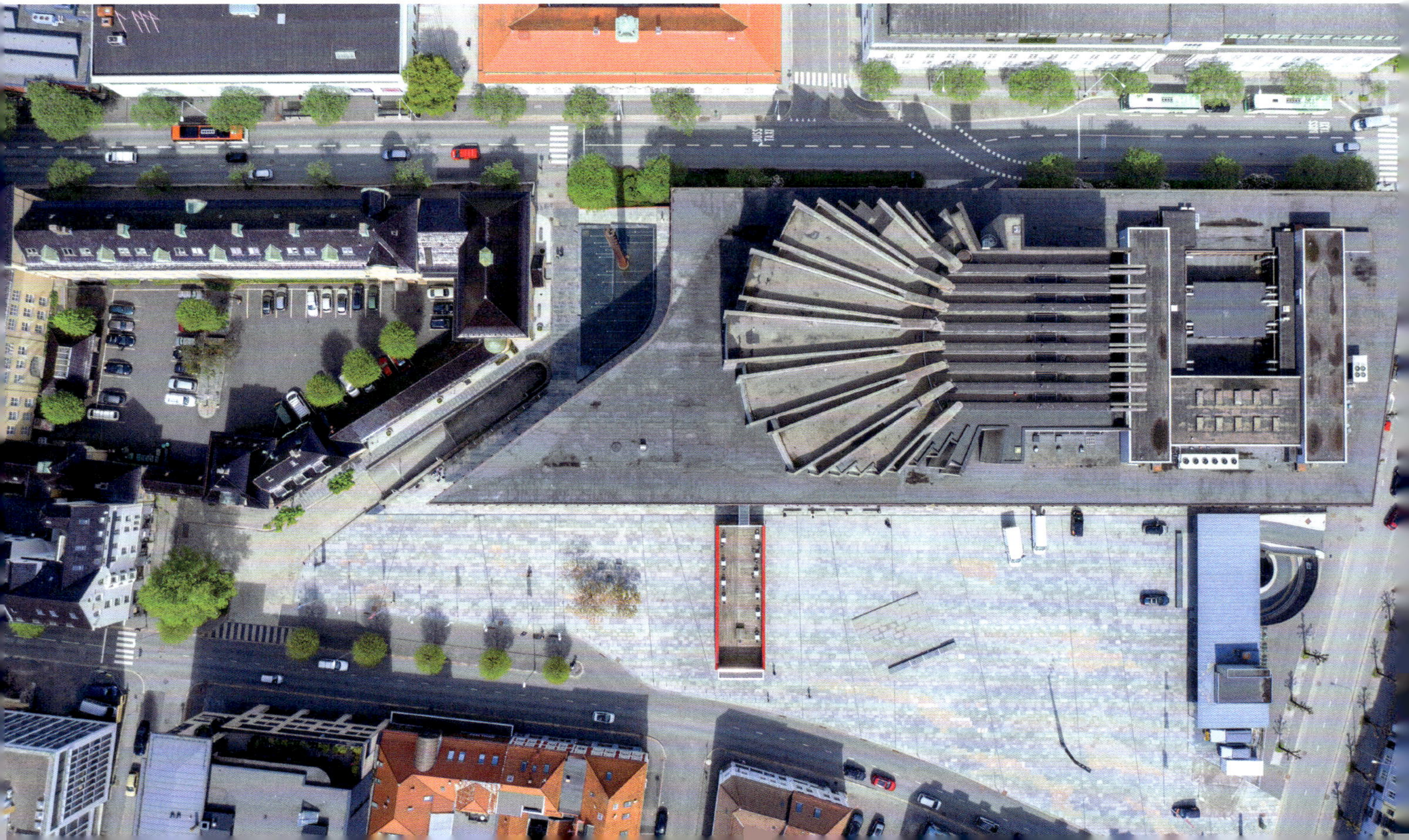

The square lies in front of Bergen's major Concert hall, **THE EDVARD GRIEG MUSIC THEATRE**.

In 2005, the board decided to give the square a total remake as a new three story garage was to be constructed below the square.

The aim was to create a public space in a highly urban context that could also function as an extended area for cultural activities.

The square should present itself as a sophisticated and attractive meeting point.

The southern part of Edvard Grieg Square are flanked by two buildings. One is the entrance area from the underground car park, the other a red pavilion containing a café, ticket office and access to the garage.

Diagonals in polished black granite are crossing the square. They are split by led-lighted glass stripes, glowing in the night.

A 17meter steel pole with a base of stainless steel provides the lighting of the this part of the square. Spotlight beamers are mounted at the top and blue light is pulsating from top of the stainless steel covering, resembling a torch.

The pole itself is designed as a guitar-neck, as a comment to the music, both inside the concert hall and on the square outside.

The Square floor is designed with the use of a plain, grey granite as base, with intarsia of both red and grey gneisses.

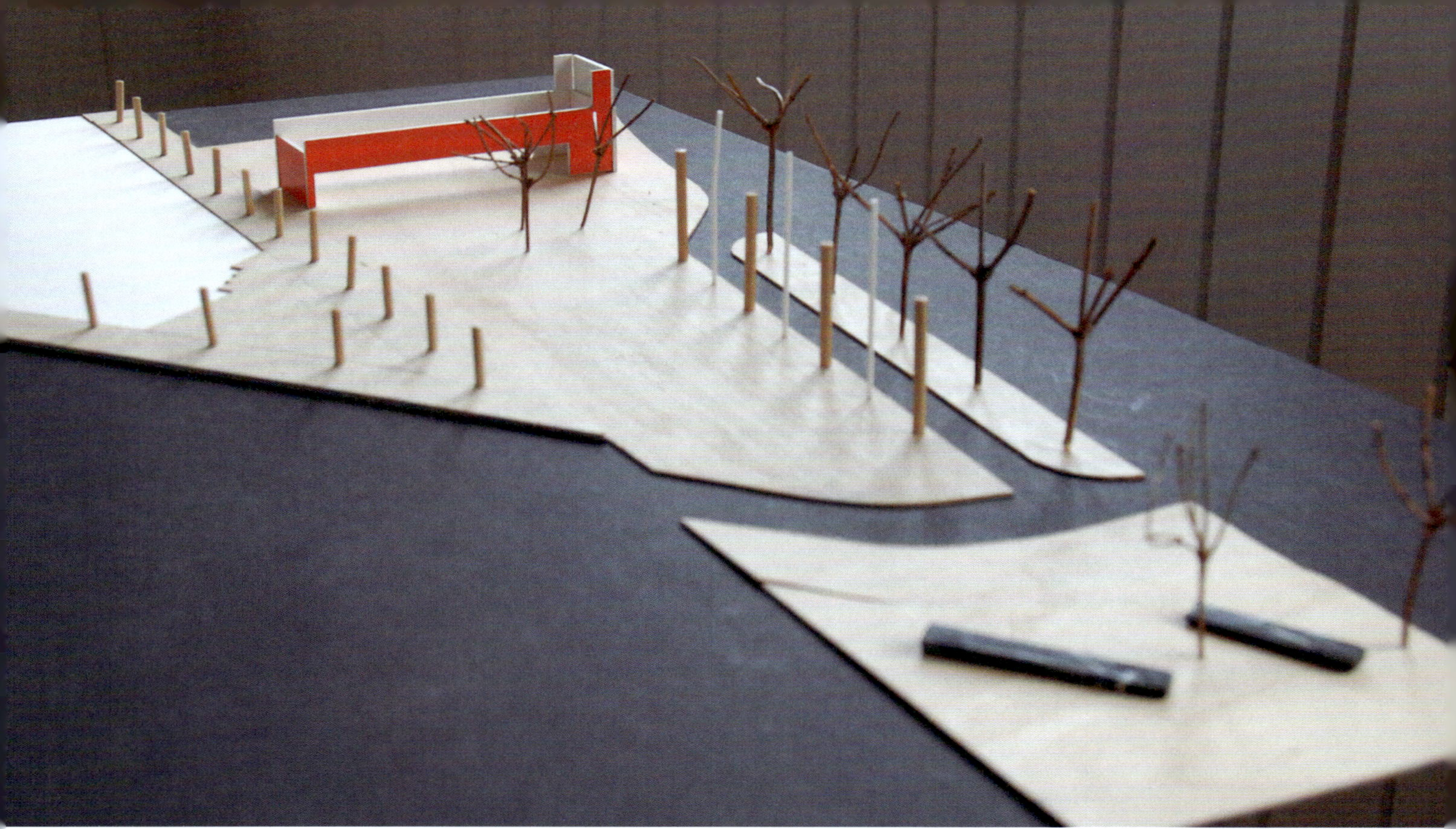

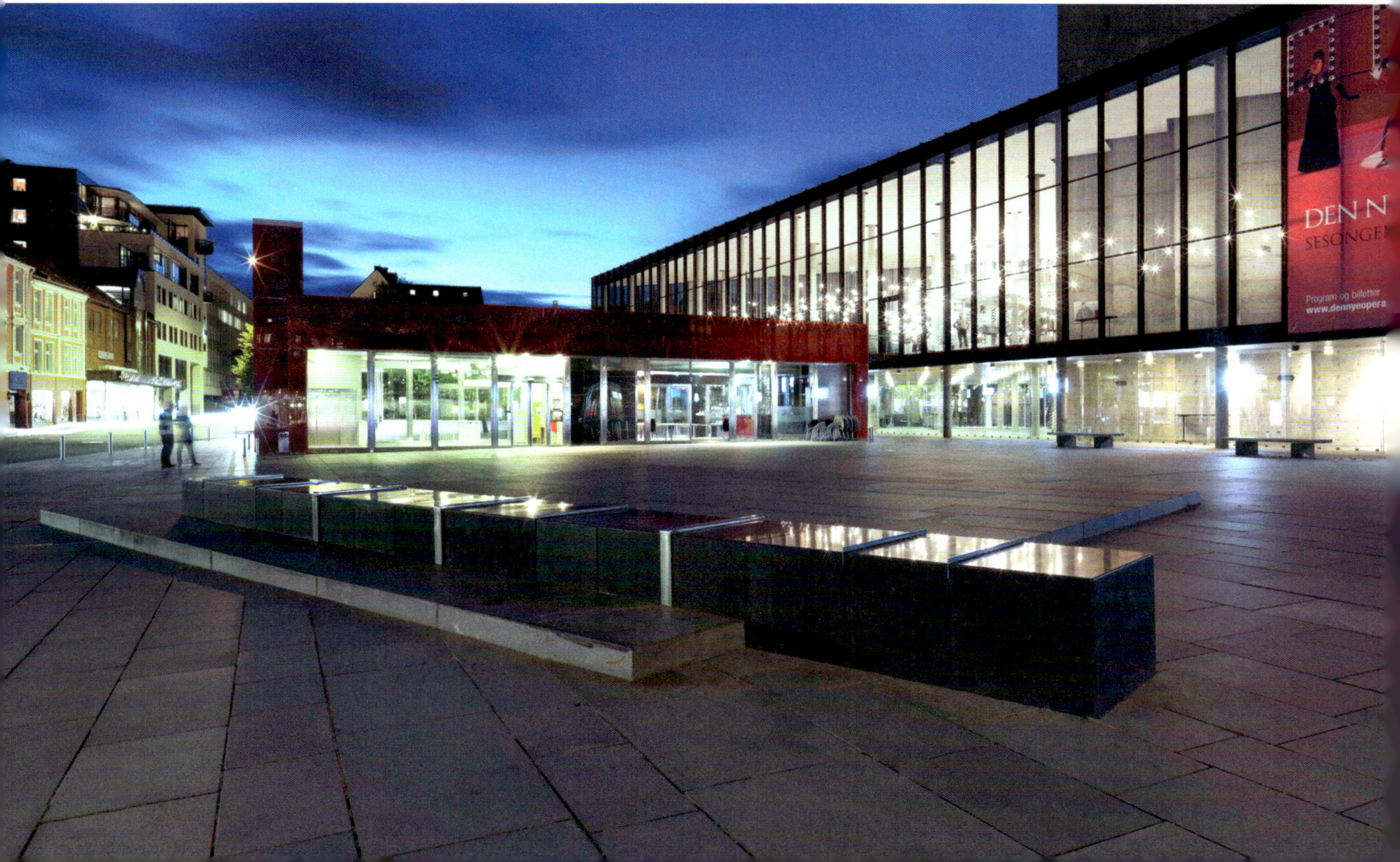

DEN N
SESONGE
Program og billetter
www.dennyeopera

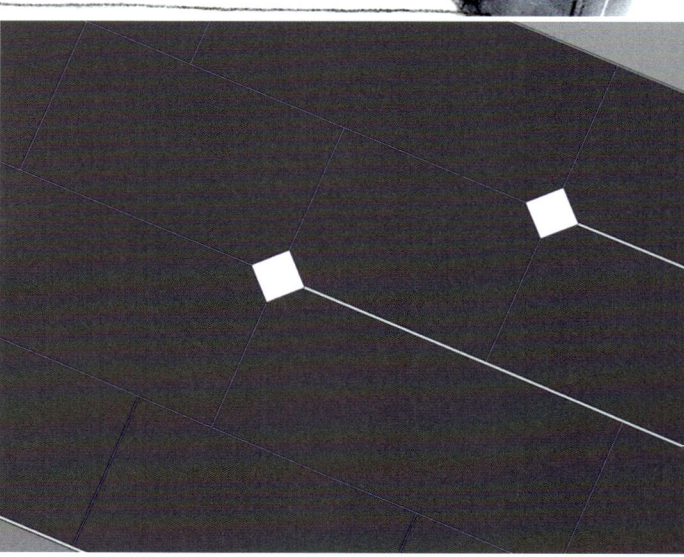

59

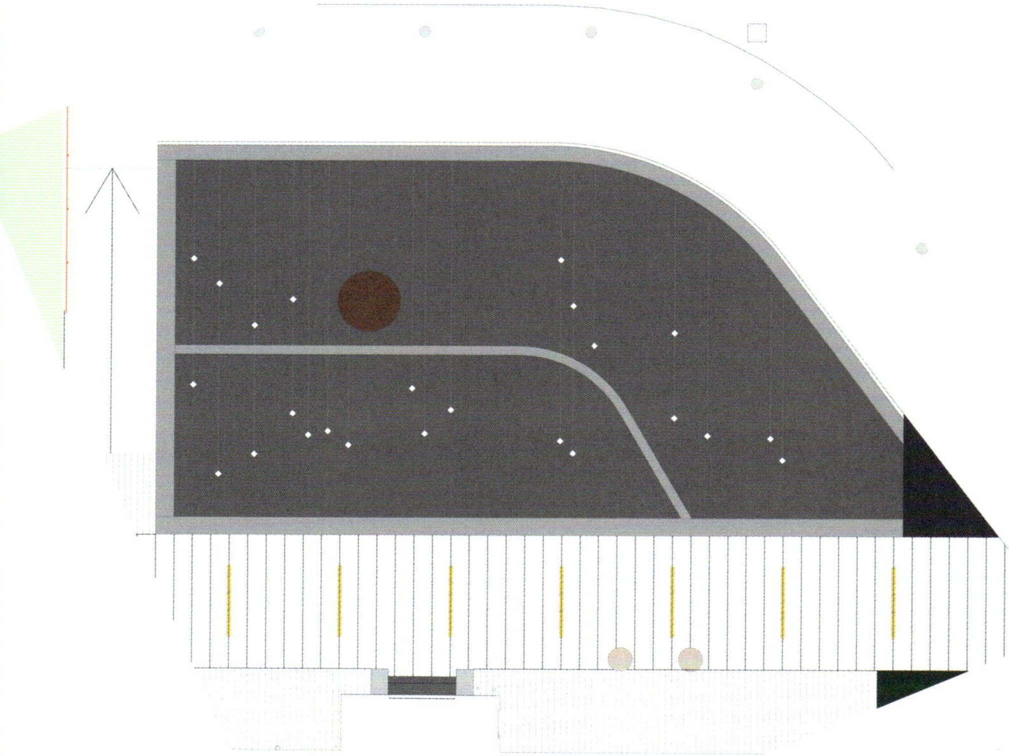

Early sketches of pool.

Light scape

HARALD STREET

HAUGESUND, NORWAY

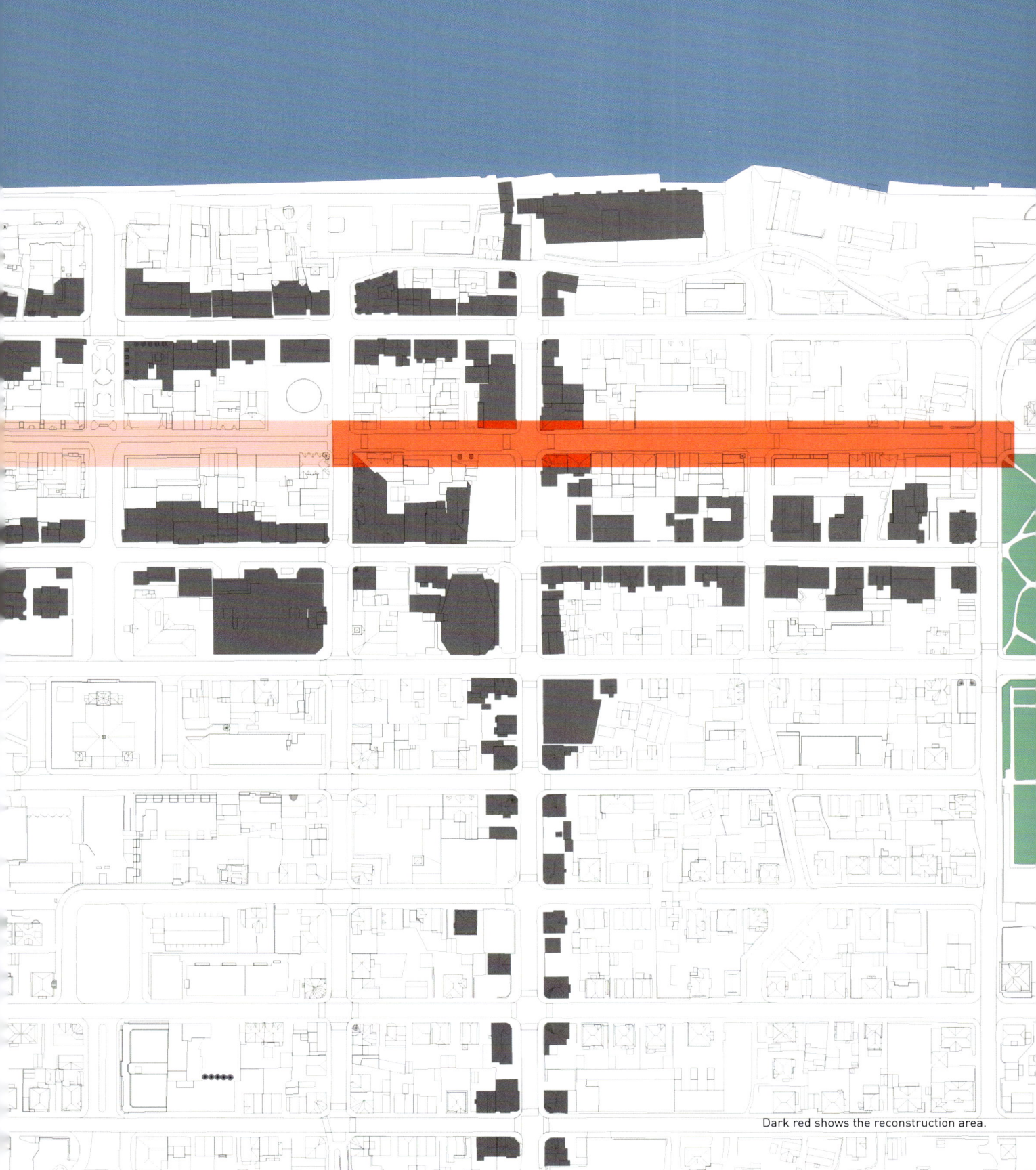

Dark red shows the reconstruction area.

Sketches streetscapes.

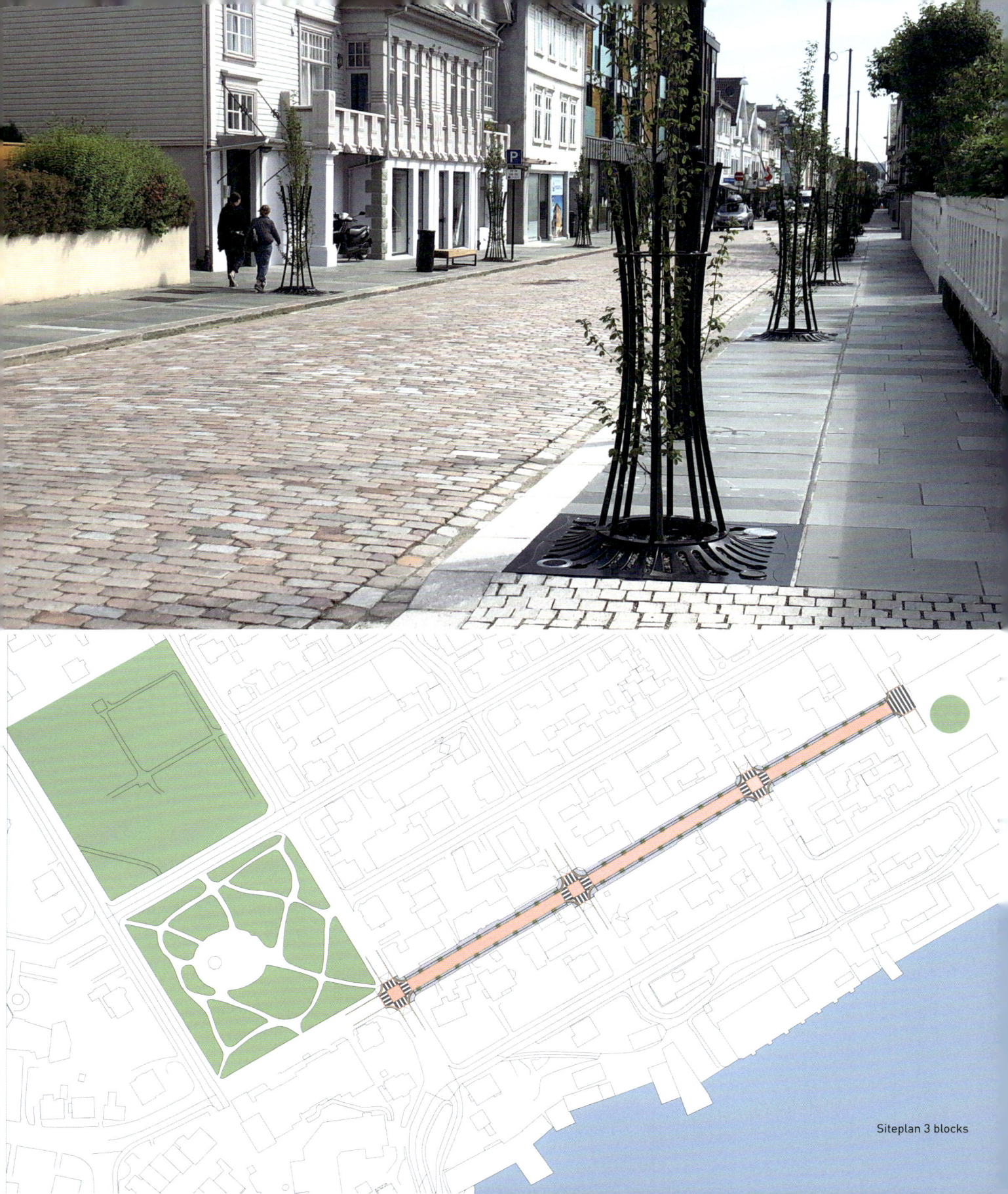

Siteplan 3 blocks

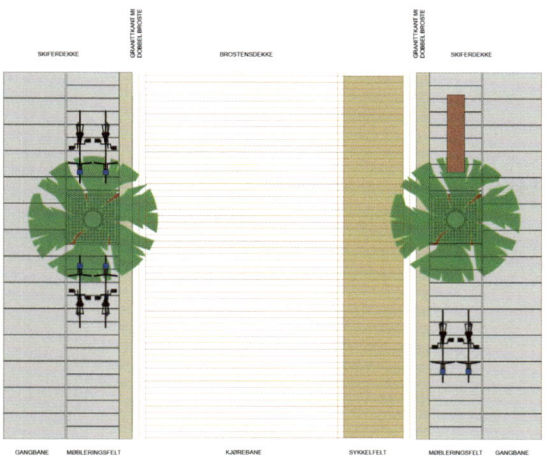

HARALD STREET is Haugesund´s shopping street. It covers about a kilometer from the Town Hall Park to the City Park. Two thirds was already established as a pedestrian zone. The remaining third should now be redesigned for mixed traffic. It should be especially facilitated for bicycling. At the same time the municipality pointed out that in the remaining three blocks it should also be used traditional materials as cobbled stone in the driving range, slates on the pavements etc. The solution for better bicycling comfort, was to saw each and every one of the cobbles in the cycle area, and afterwards flame the sawn surfaces before laying the street.

For better access in the crossings, the kerbstone was lowered to only 2 centimeters hight.

This was obtained by lifting the while area in each crossing, and in addition lower the curbs by special designed marking stones.

As a homage to the local poet Kolbein Falkeid, some quotations from his poetry in steel intarsia was laid down on the pavements on both sides of the street.

Siteplan second block.

Det fins lys
som ennå stuper gjennom
verdensrommet
uten å ha nådd jorda
annet enn som et løfte.

HARBOUR COMMON

BERGEN, NORWAY

1. Proposed solution.
2. Conceptual sketches.
3.-4. Early ideas.
5. Sketch of light poles.

1.

2.

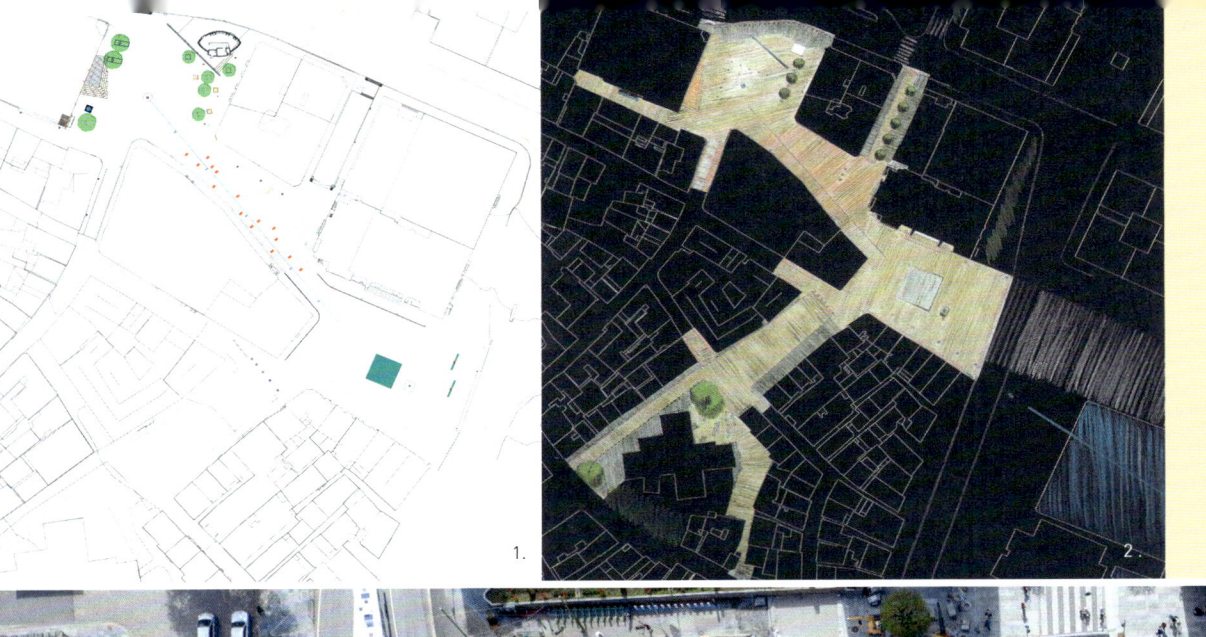

1. Siteplan
2. Sketchplan
3. Birdview

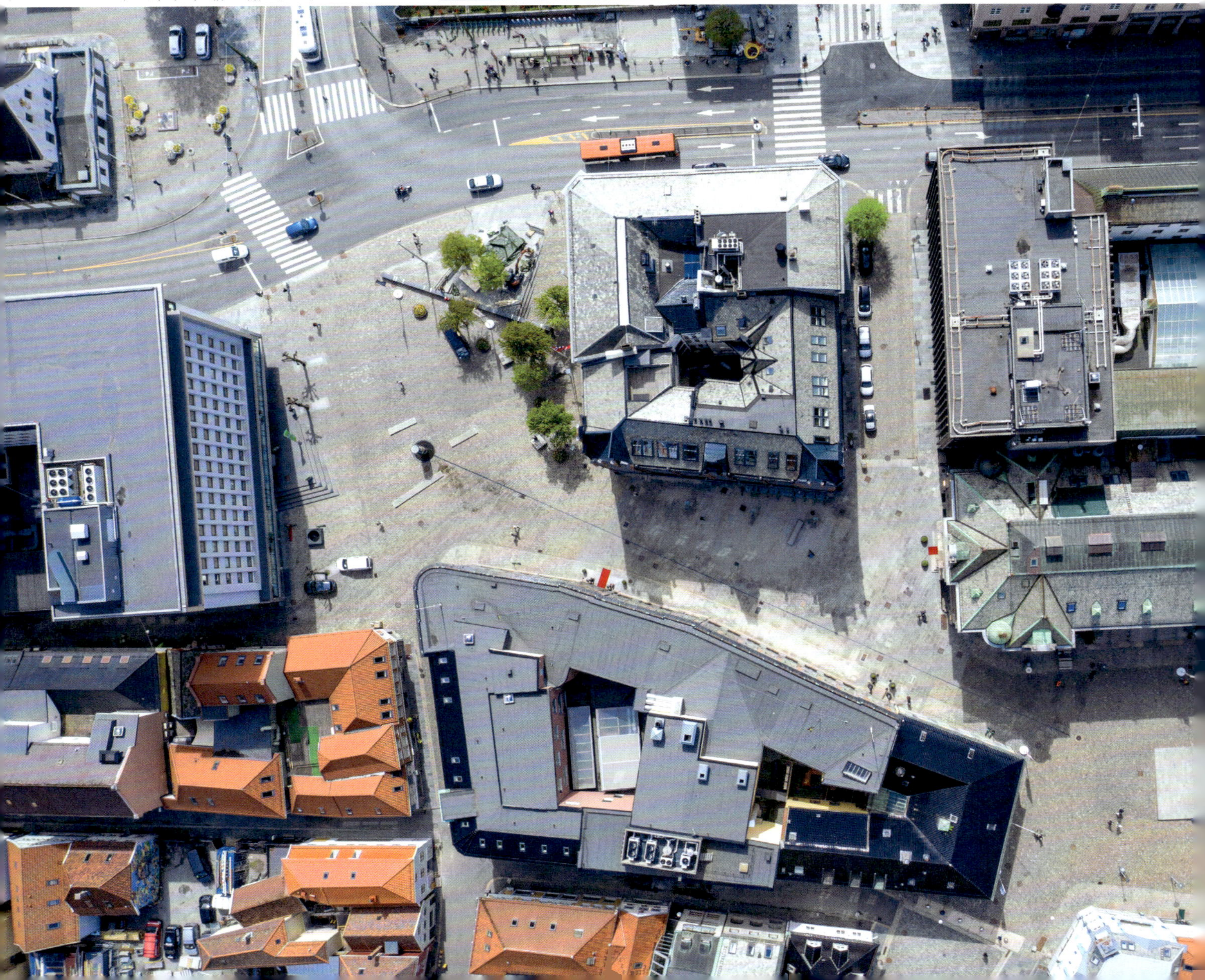

HARBOUR COMMON is an urban space with a Medieval structure. It origins from the 1300 century, and is today links the harbour to the main banking- and administrative centre in Bergen, Norway. A common of this character was made to prevent fires from spreading.

The main concept is to visualise this historical situation and the commons direct contact with the harbour.

It is really two squares connected by a widening street in the middle.

The main elements on the upper square, is a huge ceramic jar made by the artist Magne Furuholmen. It stands where the water drain starts. At random hours you may hear his composition: Steamboats of Bergen, from 3 loudspeakers in the ground around the jar.

The jar is also letting out steam, which drifts in the air and visualize the changing of winds.

On the southern side three excisting linden trees were replanted, and on the northern side stands a new planted mini-forest of Katsura tree.

The Katsura trees are lighted from fibre optic lenses inserted in huge granite slabs.

A new stair in flamed and polished black granite provides the bank entrance with a sleek ramp for wheelchair users.

Close to the stair is a well that symbolizes an ancient well from the 13th century, found while excavating the site for the bank building in the 1960ties.

It is made from two different types of black granite; one with veins and one with spots.

To separate car traffic visually from the pedestrian area a 17 metre long granite wall was constructed. The height varies from 45 to 135 cms. Inserted in the wall are several openings filled with glass. The glass is lighted from High-power LEDs beneath, but the main feature is that the cars headlights also are transmitted through the glass and down the square. Thus the pulsating traffic light becomes part of the plaza design.

The middle part between the two squares is a dark and most windy place, we decided to provide it with a higher paving quality than the two adjacent squares.

To combine the two squares, a 99 metres long ceramic drain is situated in the middle of the common.

Granite slabs covers the ground, and the artist Lasse Berntzen has created 13 specially designed slabs which contains intarsia quotations from the poet Ludvig Holberg.

A central raised box of granite functions as a bench as well as a starting point for skaters.

In the centre of the lower square a tilted platform made from soapstone is laid out. It measures 9 x 9 metres. Two benches, also made from soapstone, have incorporated glass elements, which are lighted from sunken LED-armatures.

The choice of soapstone is determined by soapstone in columns, windows-frames, door-frames and stairs in the old buildings framing the lower square.

The soapstone is soft and can easily be cut by a knife. One of the project intentions is that time can be visualised through the yearly degrading of the stone-elements.

The two squares are paved with old cobbled stones from the site. Age 200 years or more. Along the facades large, local gneiss slabs are laid out to provide better conditions for pedestrians. In the paving on the lower square, a network of blue, ceramic cobbled stones are laid out in a quadratic pattern. These ceramic elements have individual forms and their main function is to collect rainwater that provides reflecting sunlight.

Section and sketch of granite wall.

Sketches of upper part of the Common.

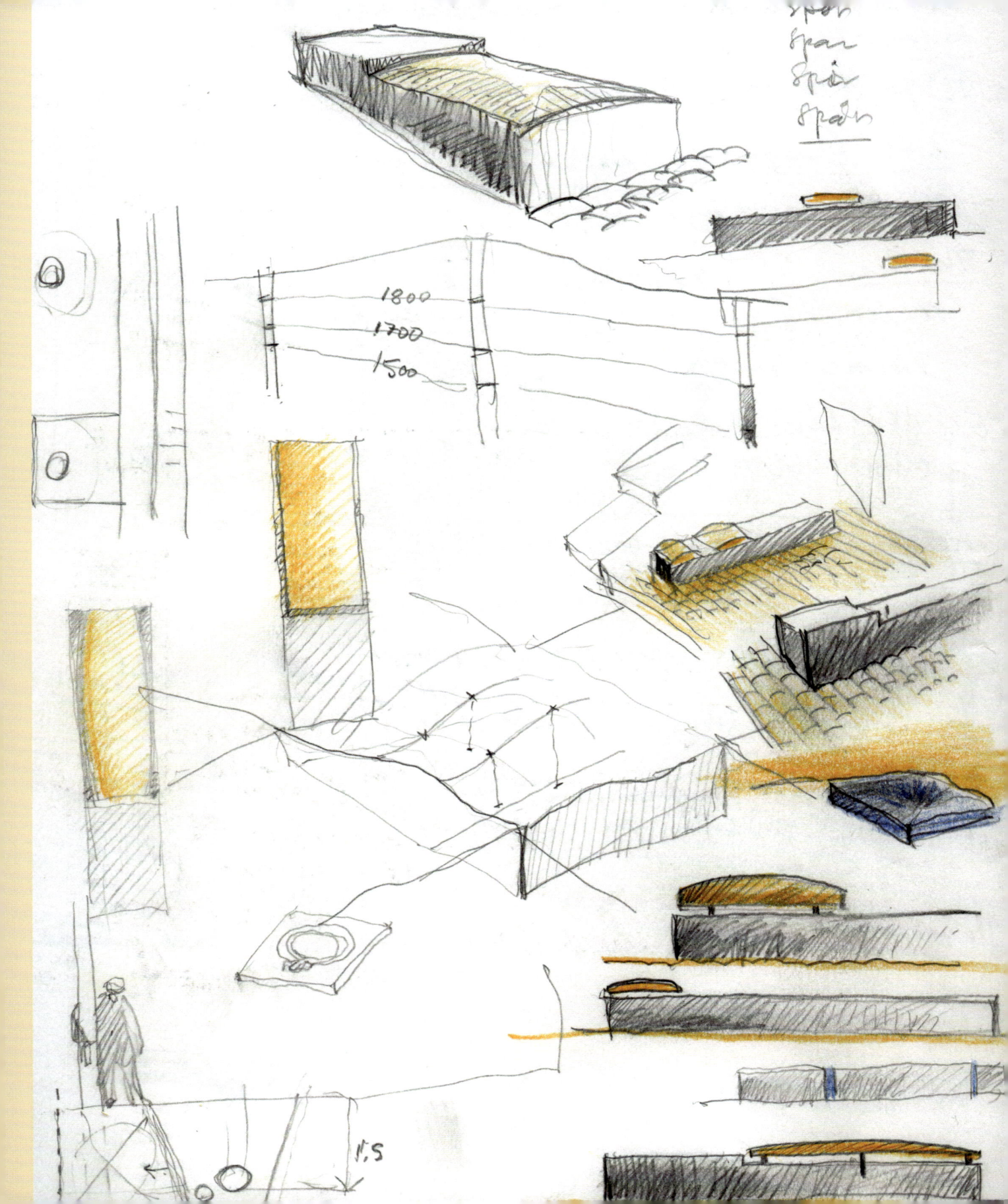

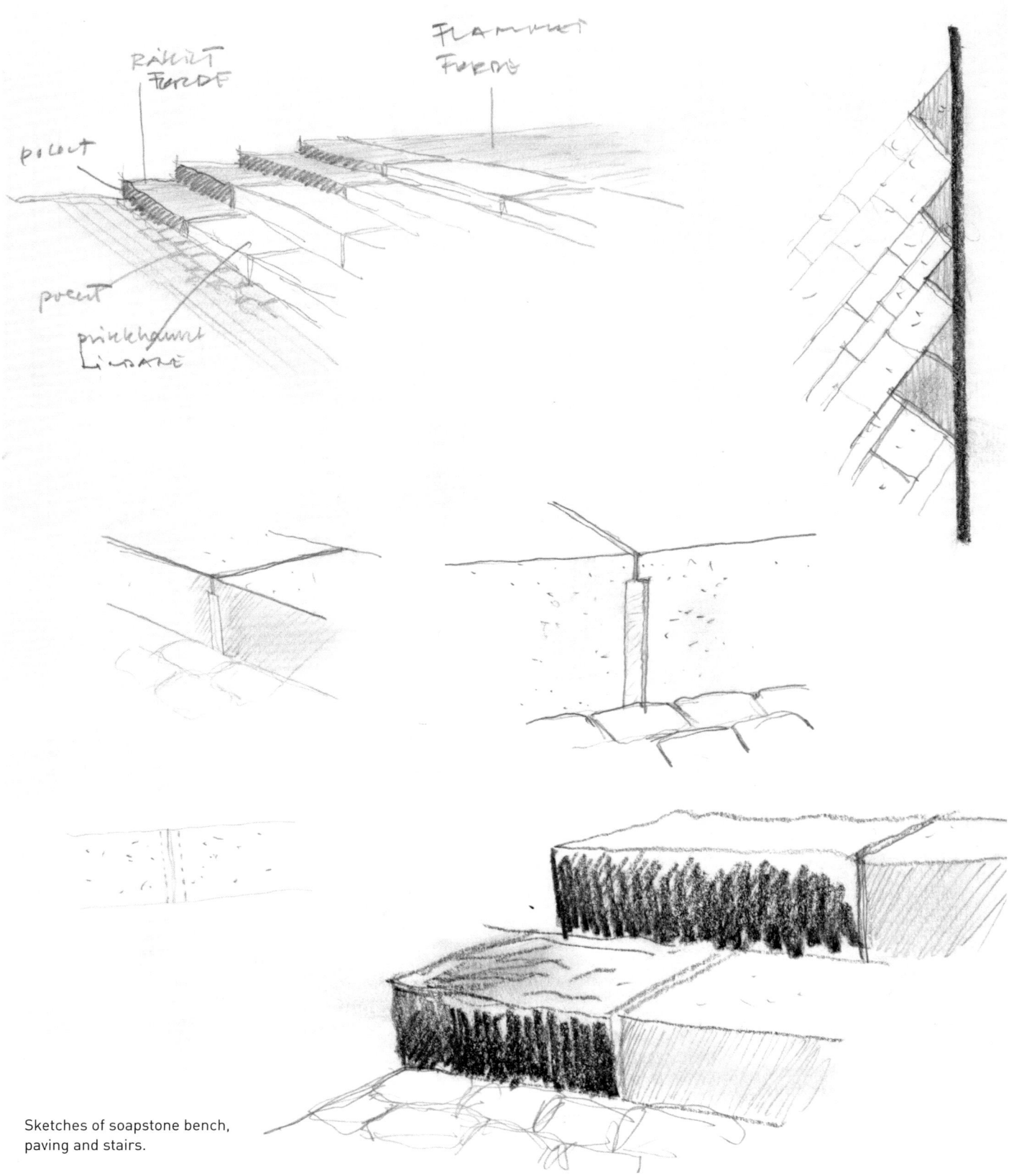

Sketches of soapstone bench, paving and stairs.

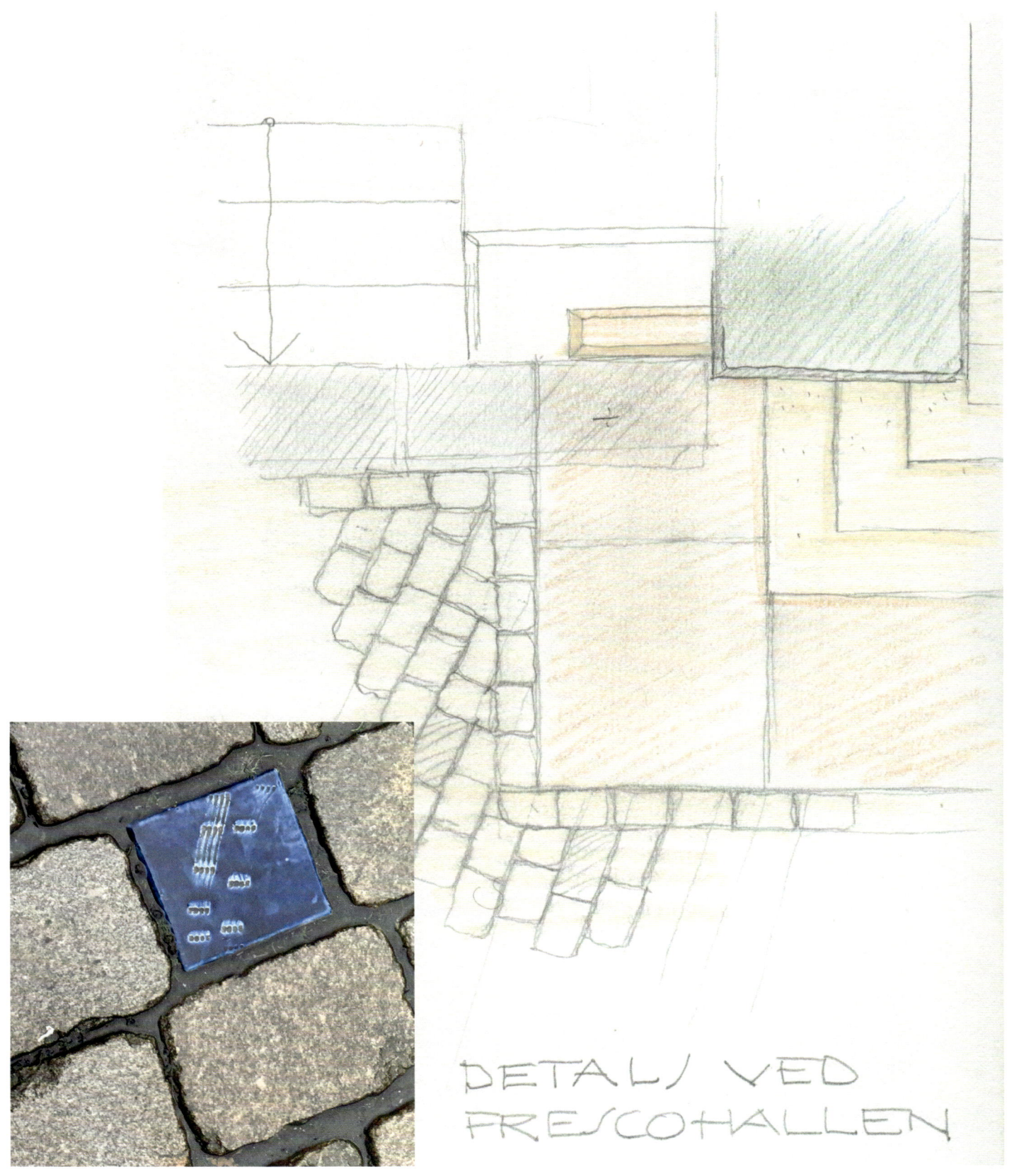

DETALJ VED FRESCOHALLEN

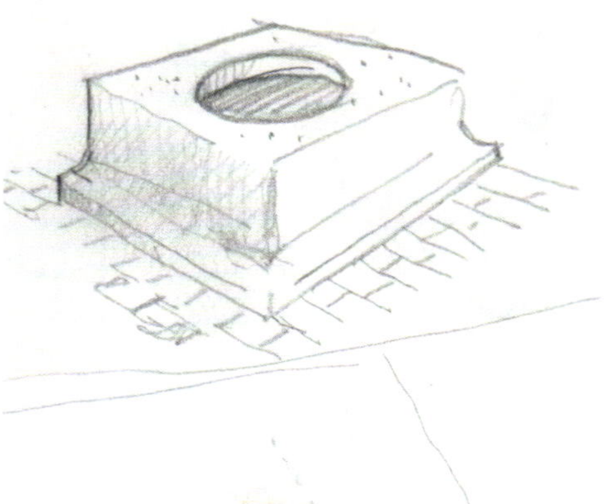

92

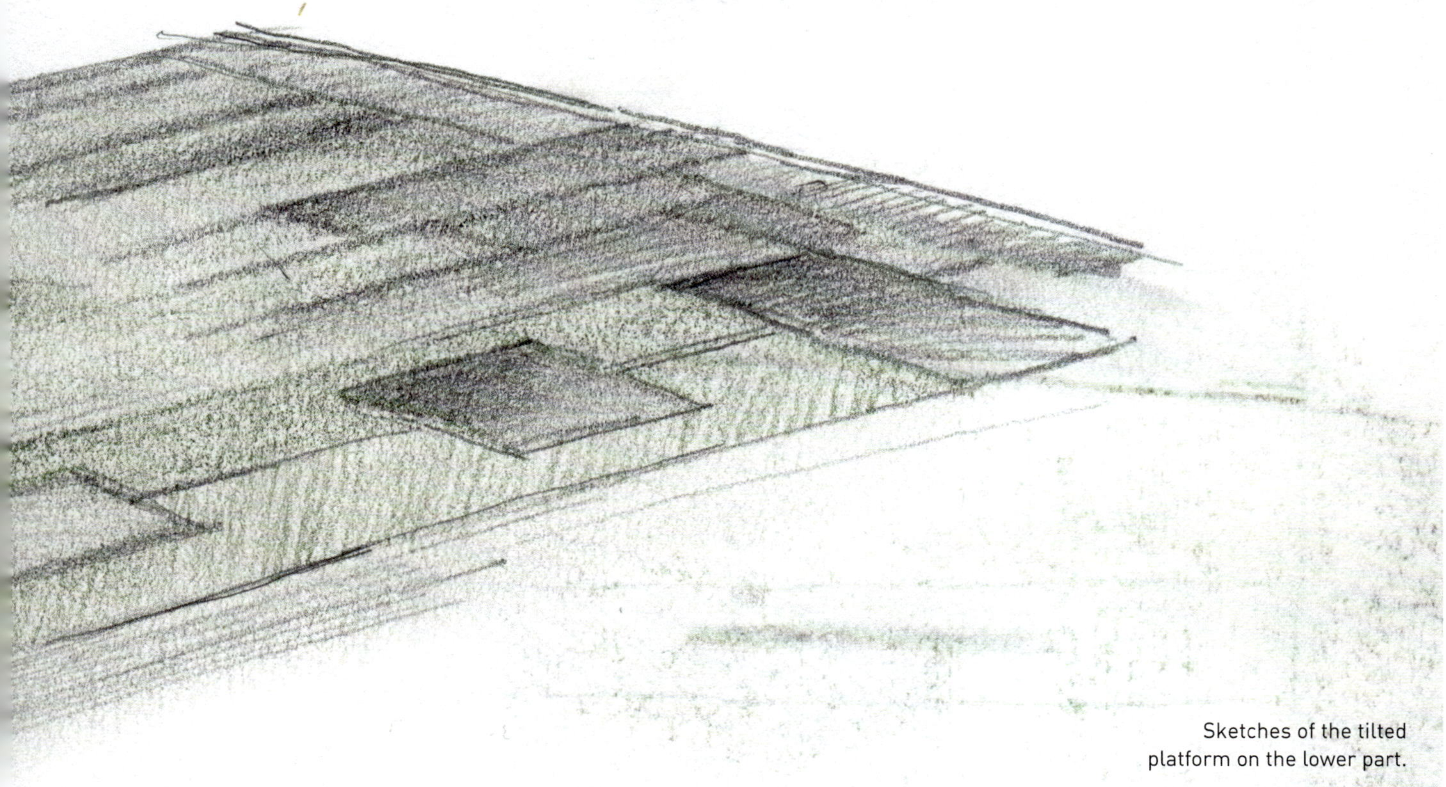

Sketches of the tilted platform on the lower part.

HARBOUR STREET

BERGEN, NORWAY

The old hanseatic city of Bergen is more than a thousand years old, and **HARBOUR STREET** is one of the oldest streets in the inner City, thus reconstructed after the fireblast of 1916.

The street is 300 metres long, 17 meters wide and slightly curved . In the eastern end the width is 22 metres.

Before the intervention, the lanes were 10/15 metres wide and the pavements 3,5 metres each. The first option was to increase the width of the pavements and thus reducing the lanes to 7 metres. To obtain this, we had to exclude car parking in the street.

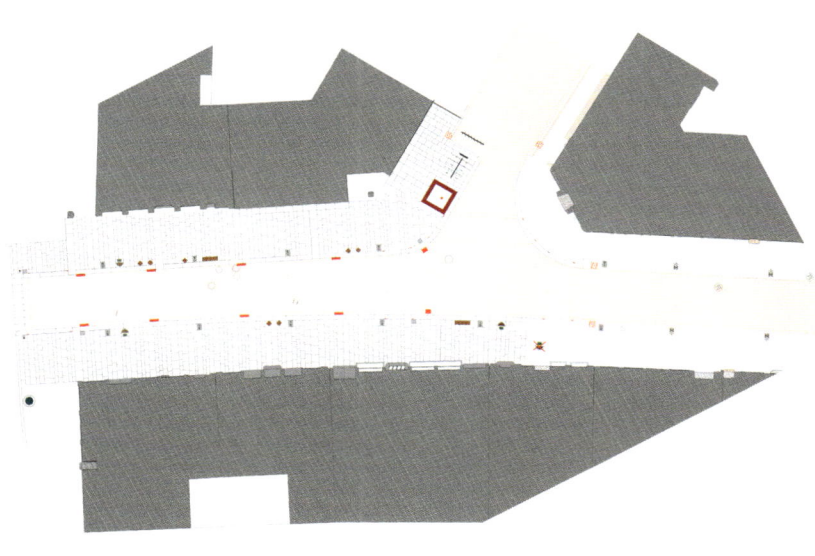

An old linden tree was on the site already, it only had to be moved a few metres. It is localized in one of the two minor squares in the street. Wrapped around this tree is a 3 x 3 m wall made in polished black granite with intarsia of local gneiss. The gneiss-elements close the gap in the joint, and create a new and contrasting pattern in the wall.

The other small square contains a new Katsura tree, and is enclosed by a huge grey granite slab. 18 holes are drilled through the granite to catch the rain. Close to this tree stands a bench totally made from stone. The sides are made from local gneiss, and the seat from a local, light pink granite. The granite bench and the tree embraced by its granite passe-partout, forms an architectural unity on this spot.

The pavements were laid out with light grey granite. To mark the old positions of the buildings prior to the fire in 1916, some 1500 ceramic bricks was hand made and filled in the joints where the former buildings´corners have been. In this way one can read some of the city´s history in the paving. These decoration are hand-made by Kari Aasen and Eli Veim.

The curbs are 60 cm wide and 30 cm thick. The visible vertical edge is only 2 cm.

The curbs have saw-marks all across the horizontal surface in order to create shadows that help poorly sighted people to orientate, and the relief gives blind people a signal of the forthcoming driving zone.

All the pedestrian crossings are made with a lowered central part in the curbs suitable for wheelchairs and trolleys. The pedestrian crossings are laid with black and white cobbled stones. The main pedestrian crossing is laid with massive, black granite with flamed surface. They are 60 cm wide and 40 cm deep. In the spaces between there is used white cobbled stones.

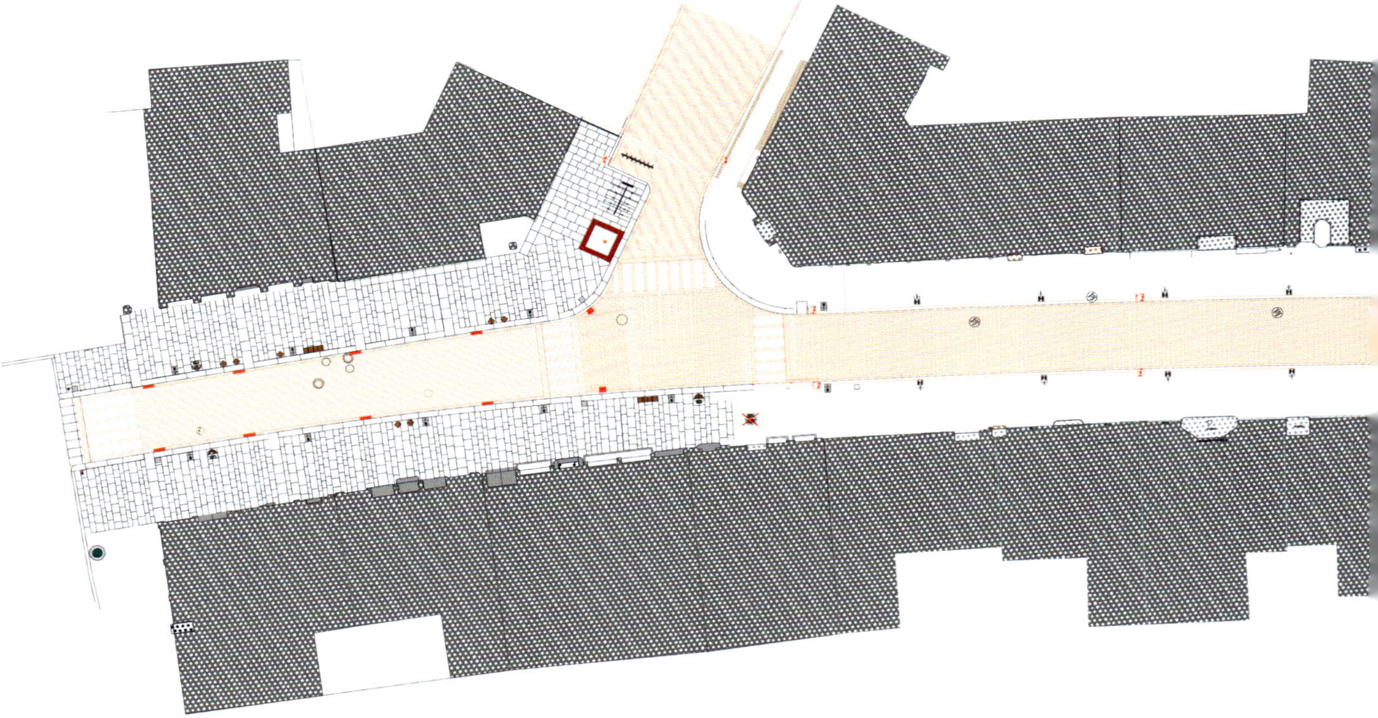

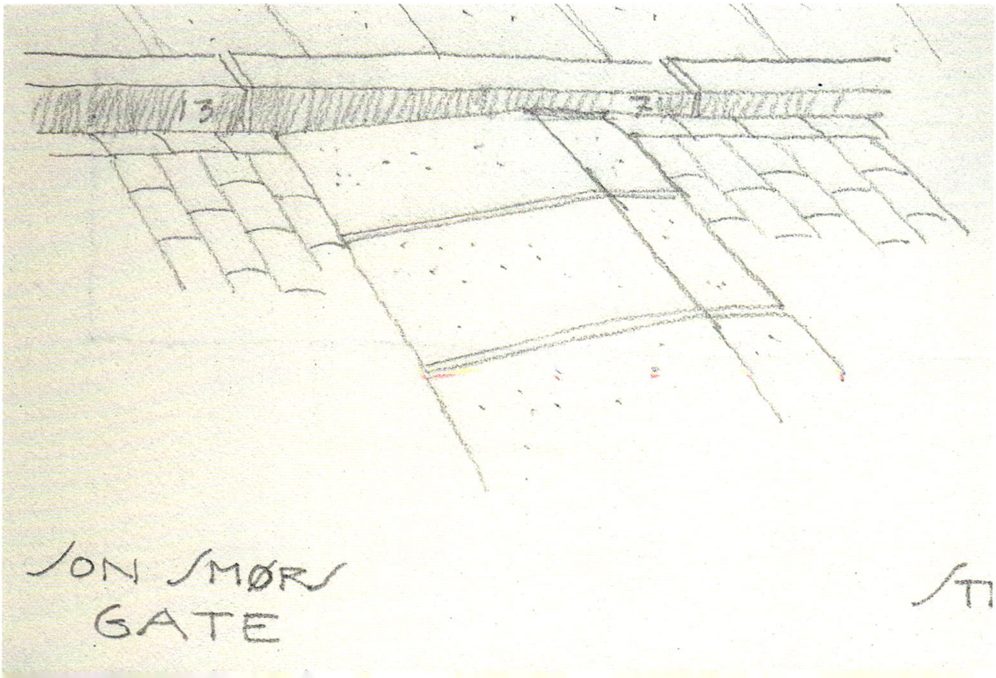
13
7
JON SMØRS
GATE

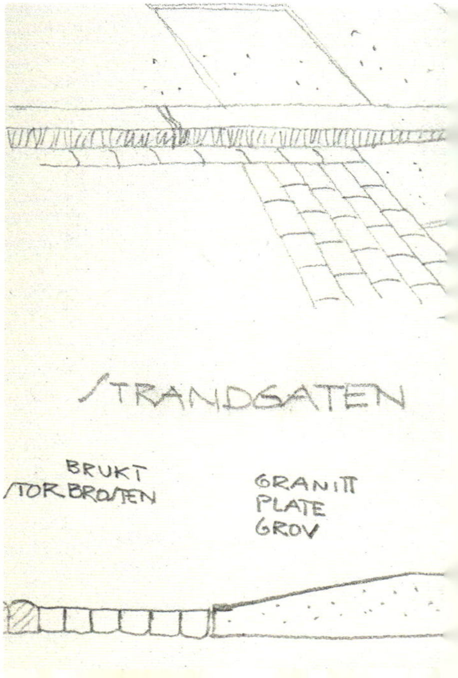
STRANDGATEN
BRUKT
STOR.BROSTEN
GRANITT
PLATE
GROV

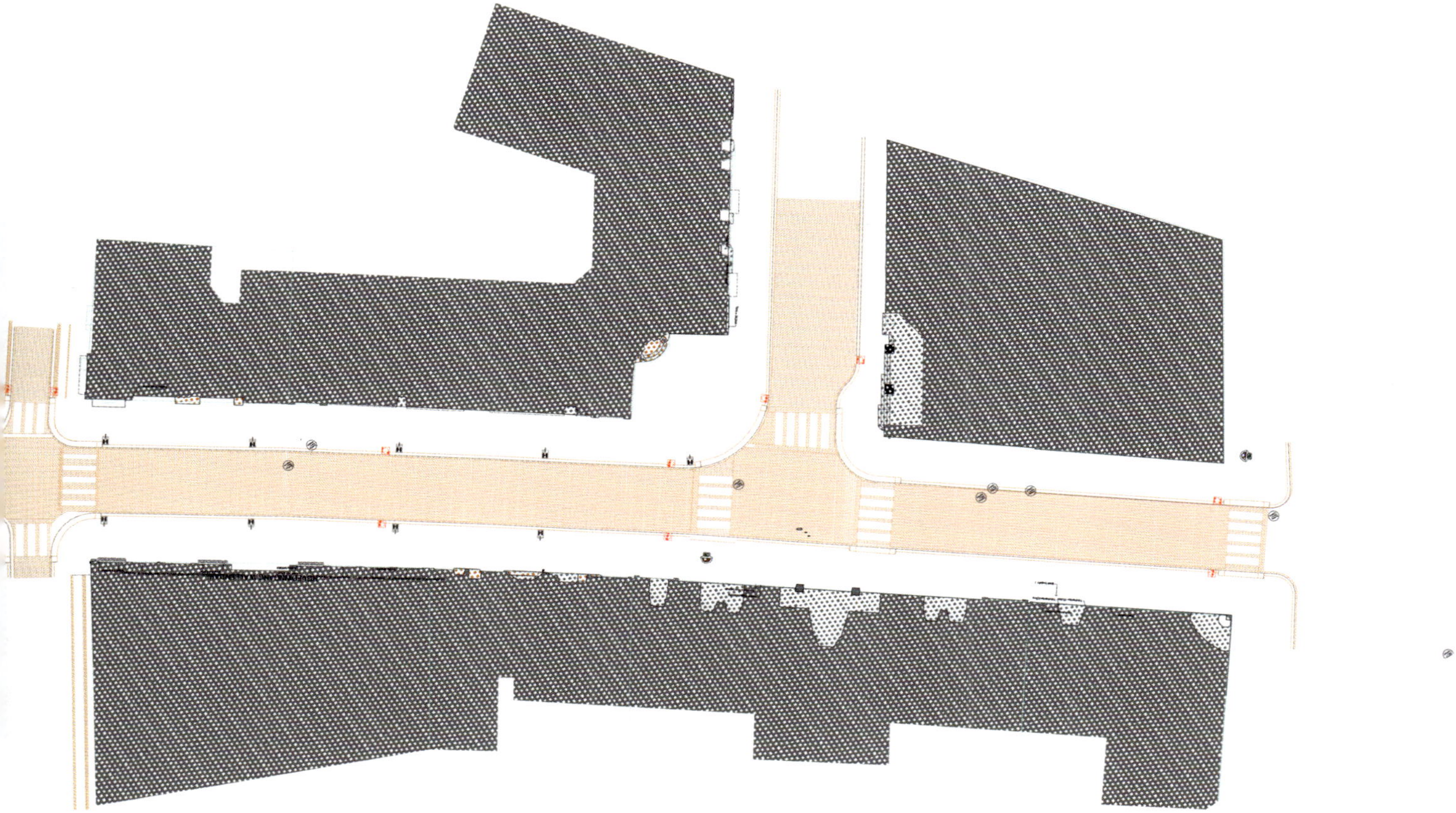

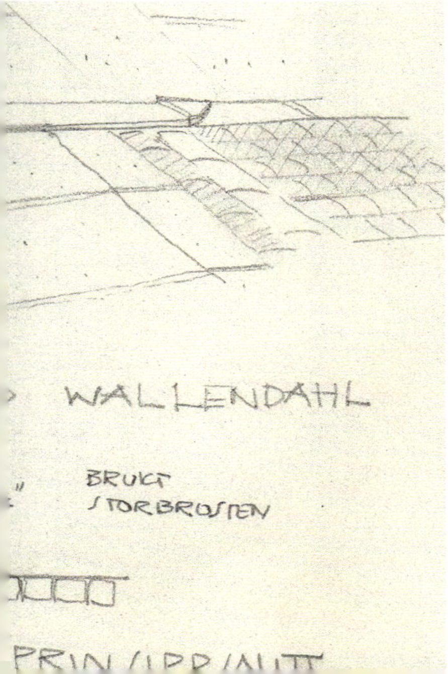
WALLENDAHL
BRUKT
STORBROSTEN

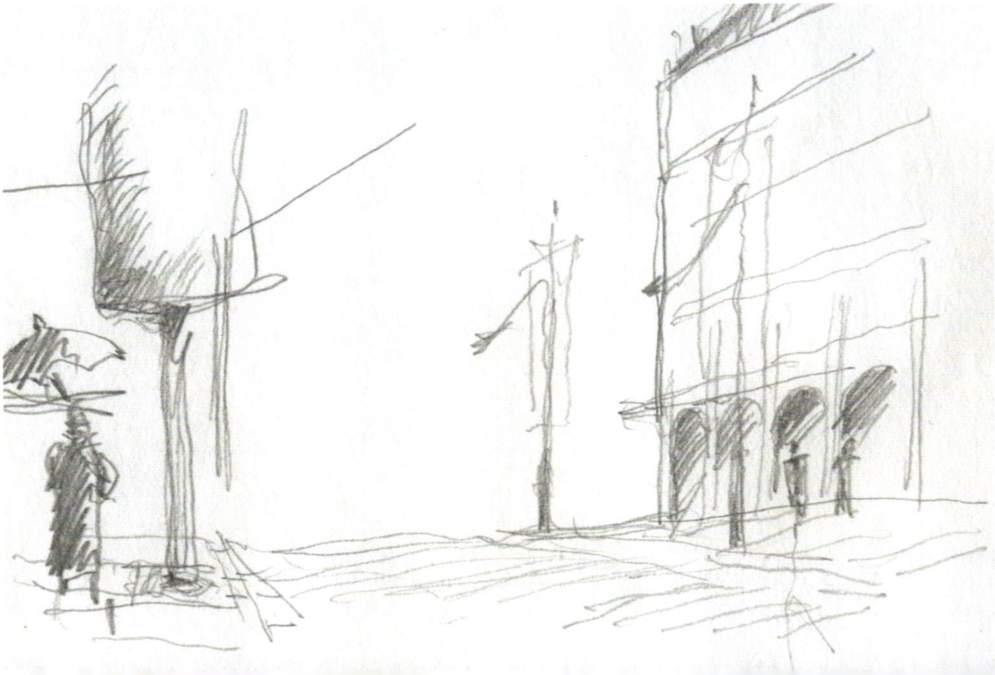

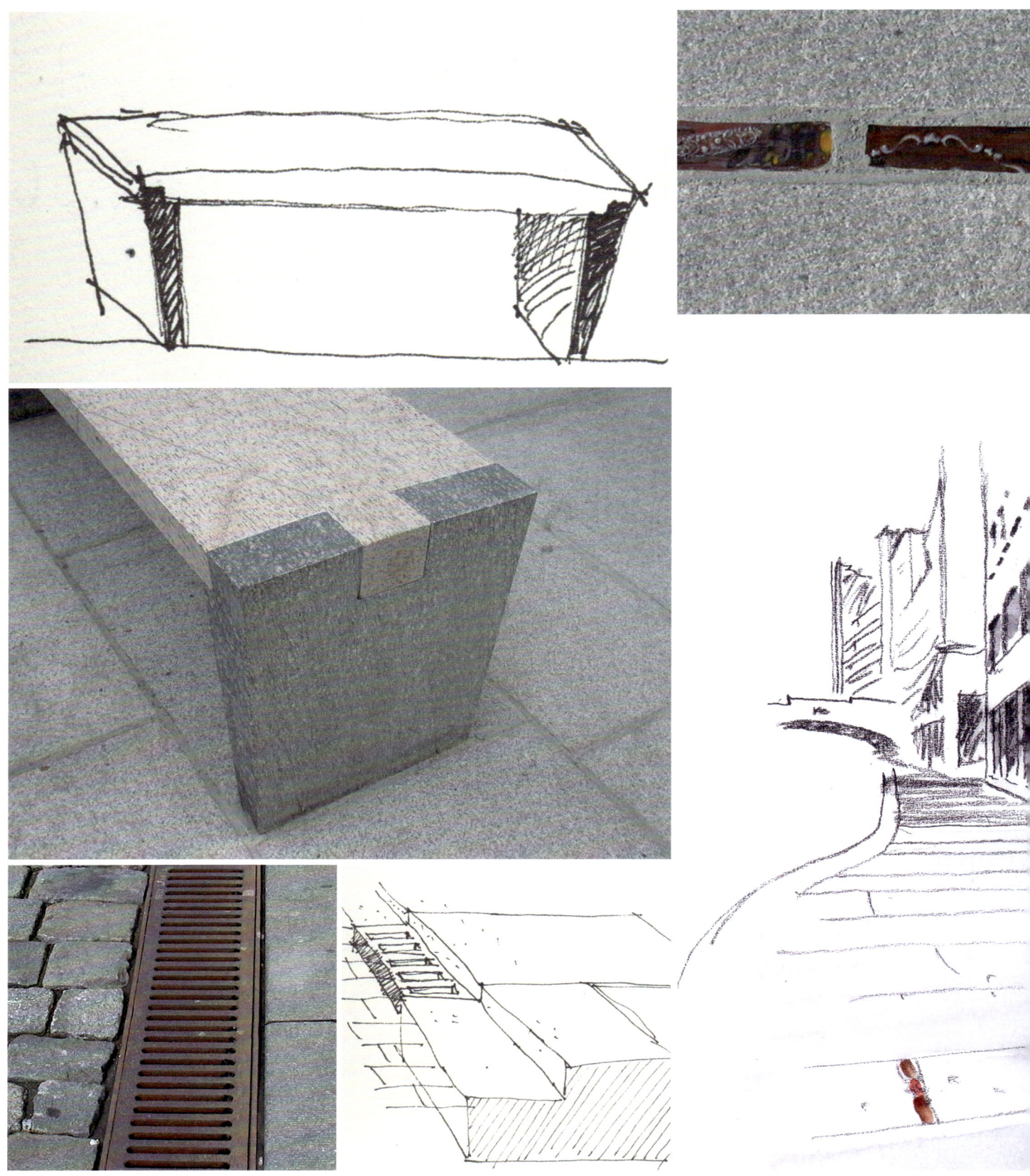

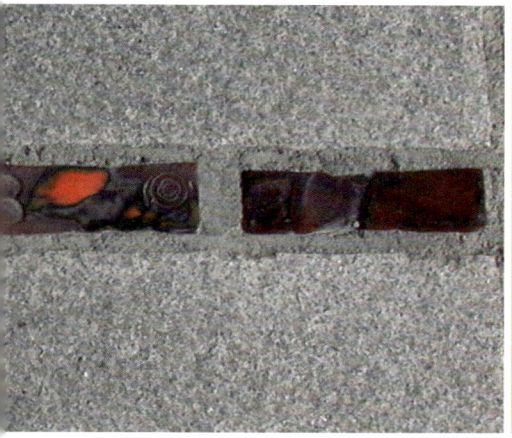

Most of the street is laid with local cobbled stones from the street. They were taken into depot and reused in the new driving zone. In the broadest part are used cobbled stones with a large, flat surface was used to create a flat area.

The street have almost no slope at all, so I designed storm drains in cast iron to take away the rain.

The lighting-system consists of 8 metre light poles with floodlights that provide direct light for the driving zone as well as reflective light for the pavements. The reflection is made by tilted and sanded glass above the floodlight.

Litter bins were custom designed for this project, made from 6 mm galvanised and painted steel.

KING OSCAR STREET

BERGEN, NORWAY

Masterplan

KING OSCAR STREET is probably the oldest street in Bergen. It had been neglected for many years before the reconstruction started in 2016. Because of the historic development, the street had a lot of old stone material which was reused in the new design.

The design was based on the shared space philosophy where all have the same right to use the whole street. Thus drivers have to drive carefully with low speed.

With 2-3 meters of rain each year, Bergen is used to handle the surface water. On the streets lower part the old curb stones were reused in a 5 centimeter curb.

On the upper large granite slabs were laid out at the same level as the cobbled area in the middle.

Bergen Cathedral is situated in the northern part of the street. The church hill was reduced through the centuries to a chaos of rose beds, traffic lights and traffic signs.

The square was redesigned as an open space cobbled with reused, white cobbles and nothing more.

This gave the cathedral back its dominating place in the city scape. Special light design illuminates the church tower in the night.

Adjacent to the cathedral lies the cathedral school, a gymnasium that dates to the early 1100.

A new entrance to the school was coordinated with the redesigning av the street.

A serie of graphic design by the artist Lasse Berntzen is Incorporated in the granite and slate-paving.

The design is inspired by historical events along the street the past thousand years.

Principal section

1.
2.

A serie of graphic design by the artist Lasse Berntzen is Incorporated in the granite and slate-paving.

the design is inspired by historical events along the street the past thousand years.

1. Perspective drawing visualising the new intervention.
2. King Oscar Street after intervention.
3. The cathedral and gymnasium in front.
4. 16 cm granite slabs in combination with large cobbled stones.
5. One of Lasse Berntzen´s intarsia in the pavements.

The Cathedral Square

1.

2.

1. Before intervention.
2. After intervention.
3. After intervention.
4. Perspective drawing visualising the new intervention.

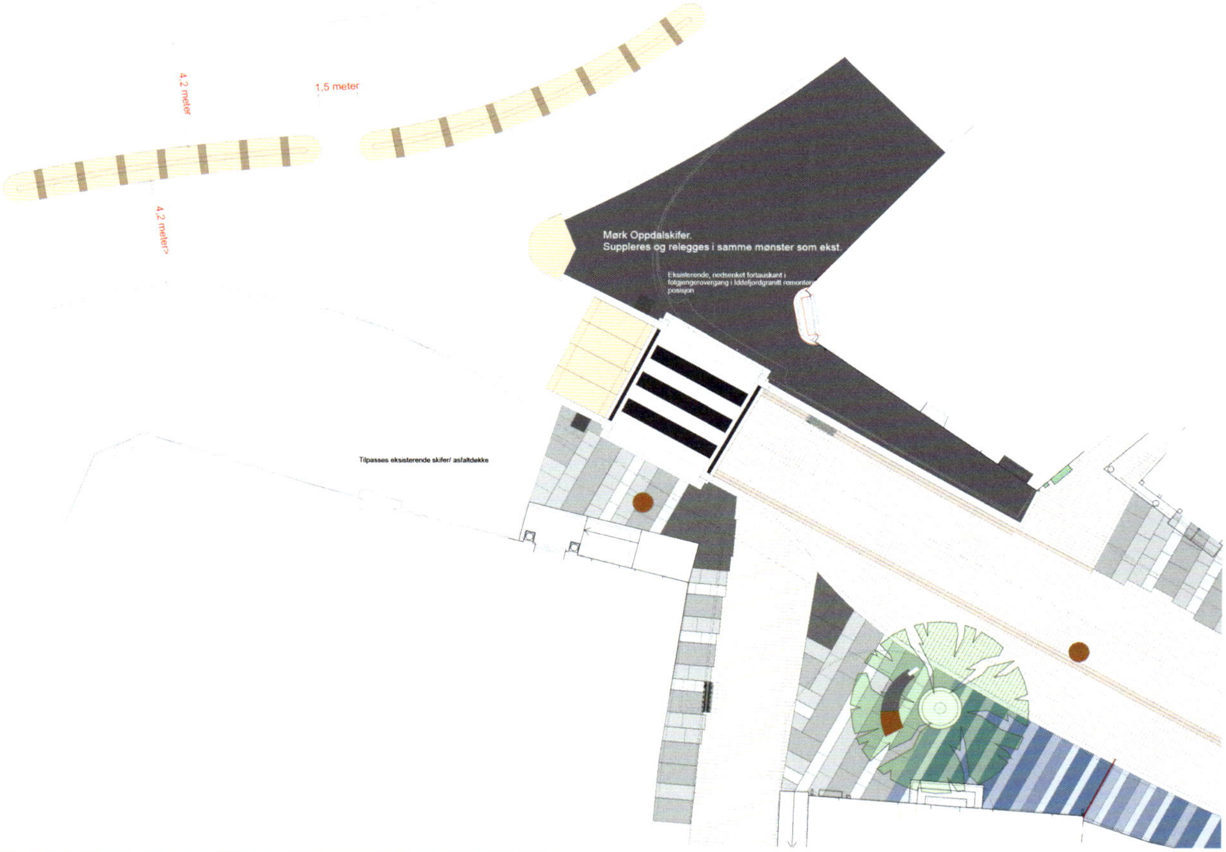

1.

1. Striped granite separates the trafic.

2. Bench in polished granite and Carrara marble.

3. Sketch of the trafic separator.

4. Sketch of bench.

5. Final drawing of bench.

Balsam
OROCCANOIL

2.

550
627
527
510
680
500
R=2 100
2 870
+ 5,0
+ 4,0
+ 2,5
+ 2,5
800
±0,00
±0,00
250
125
125
R=1 600
2 070
Granite G654 with broad veins
500
50
571
Granite G654 with broad veins
White marble, Carrra Bianca ie

5.

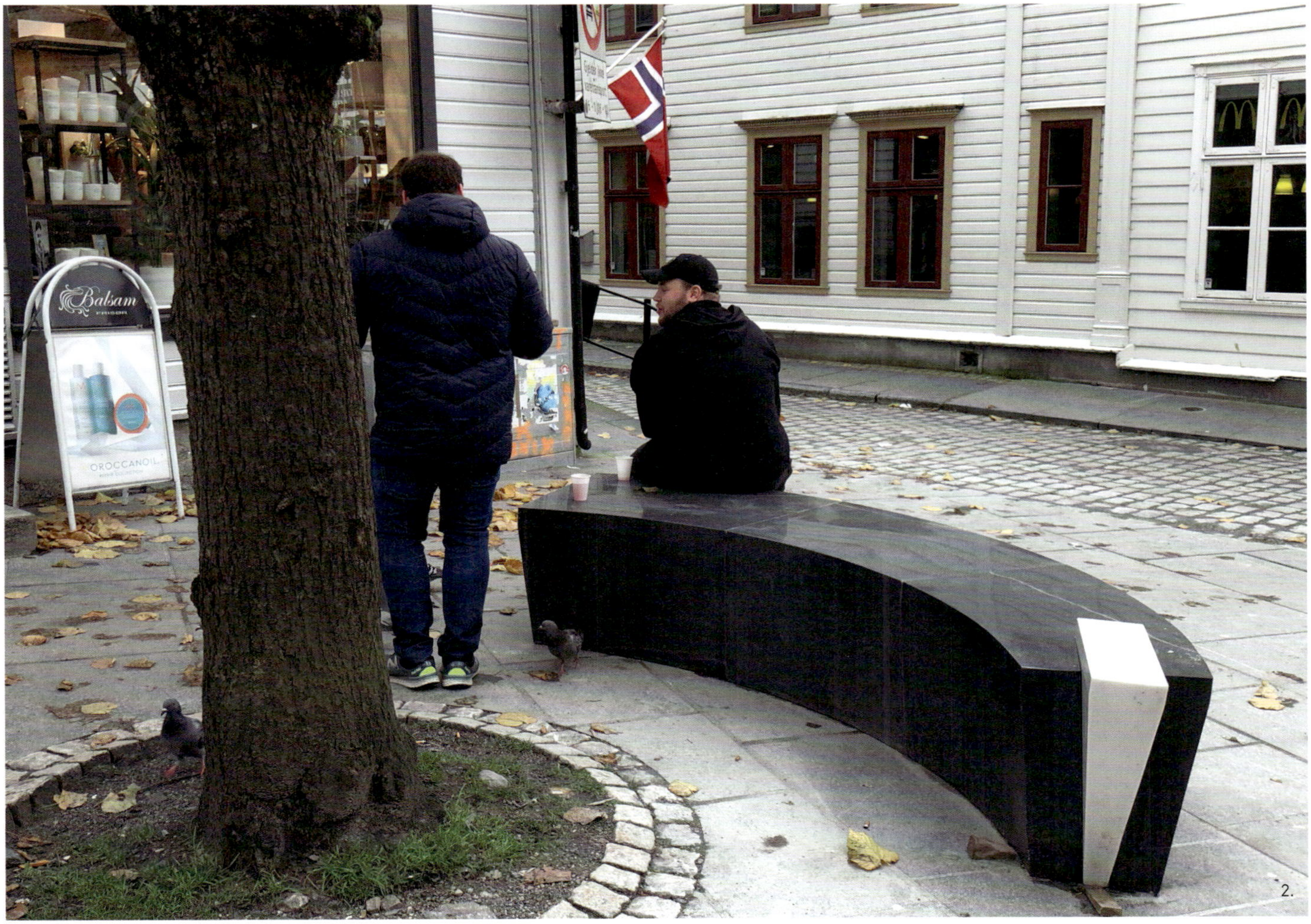

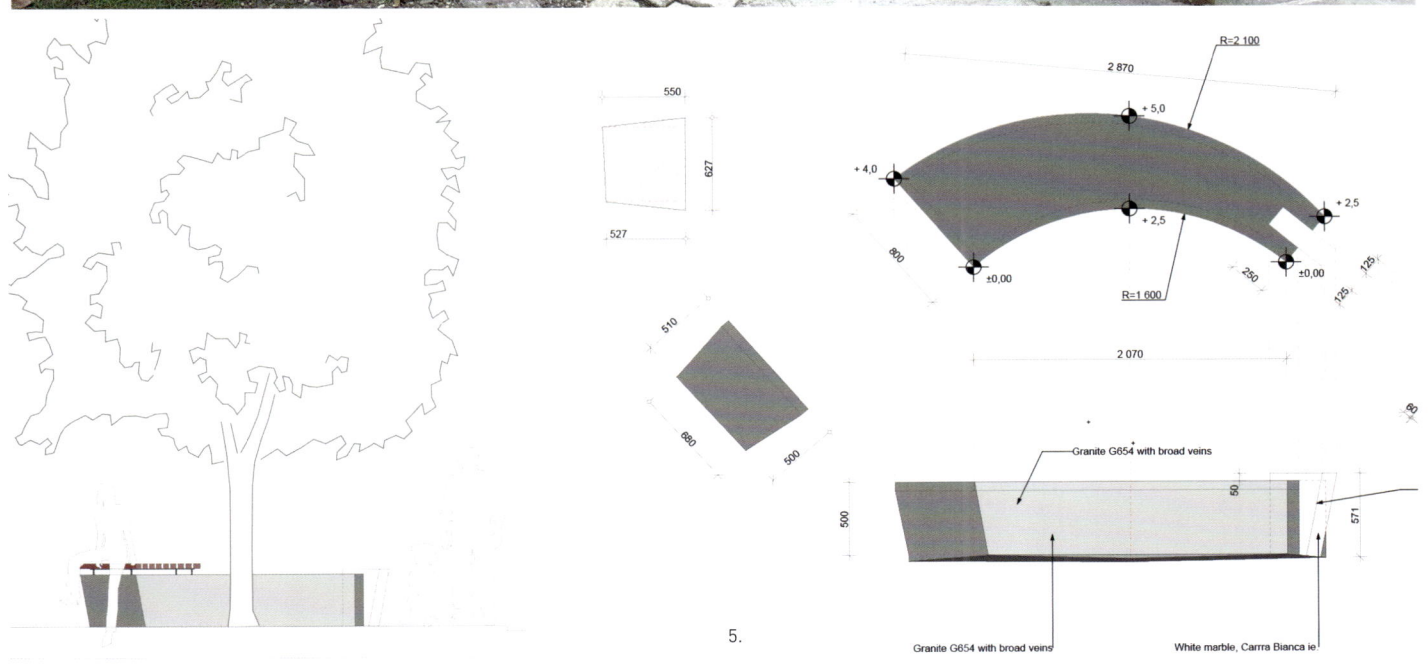

LONG STREET

SANDNES, NORWAY

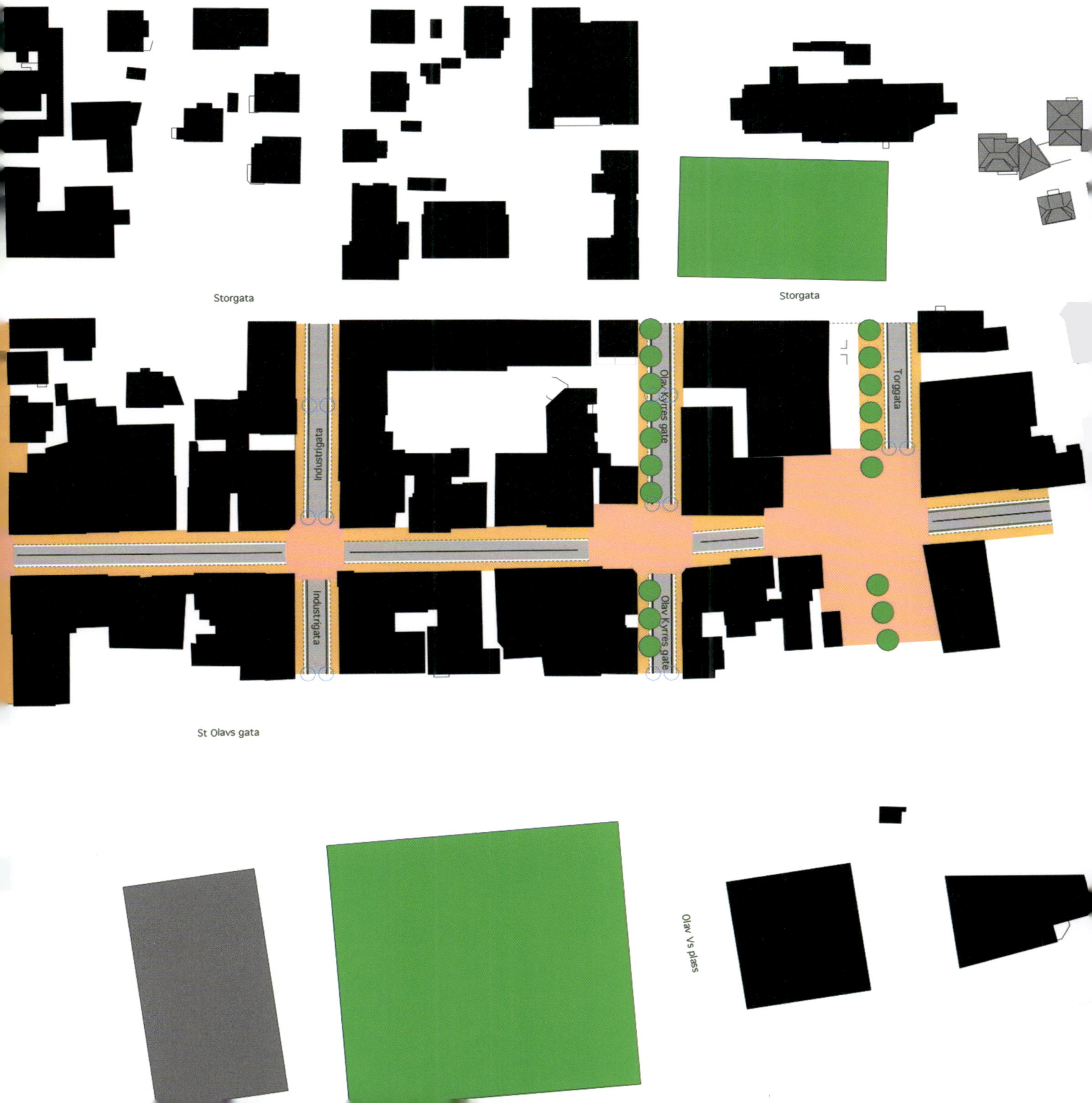

Storgata
Storgata
Industrigata
Olav Kyrres gate
Torggata
Industrigata
Olav Kyrres gate
St Olavs gata
Olav Vs plass

LONG STREET is the longest street in the centre of Sandnes. Thus the name. It stretches for more than seven blocks and is the citys main commercial district.

It was converted to pedestrian area in the 1980s. Thirty years later it was in need for reconstruction.

The city started a massive program with new drainage system, plan for greenery and lighting, as well as an artistic programme for the whole street.

Altogether five artists were invited to work in the project. Their work spans from traditional sculpture, via ceramic, cast iron and glass intarsia to light sculptures. All artistic work was wholly integrated in the design.

The main paving design distinguish the crossings from the street.

The crossings have their individually design, while the street has a common design. With a lowered drain in the middle and area for commercial activity closest to the adjacent buildings. With a pedestrian zone on both sides of the drain, all in granite slabs. The border between the pedestrian area and the commercial zone was marked with a 60 centimeter broad line with pineapple chopped surface. This is both a tactile and visual element that the poorly sighted easily can follow.

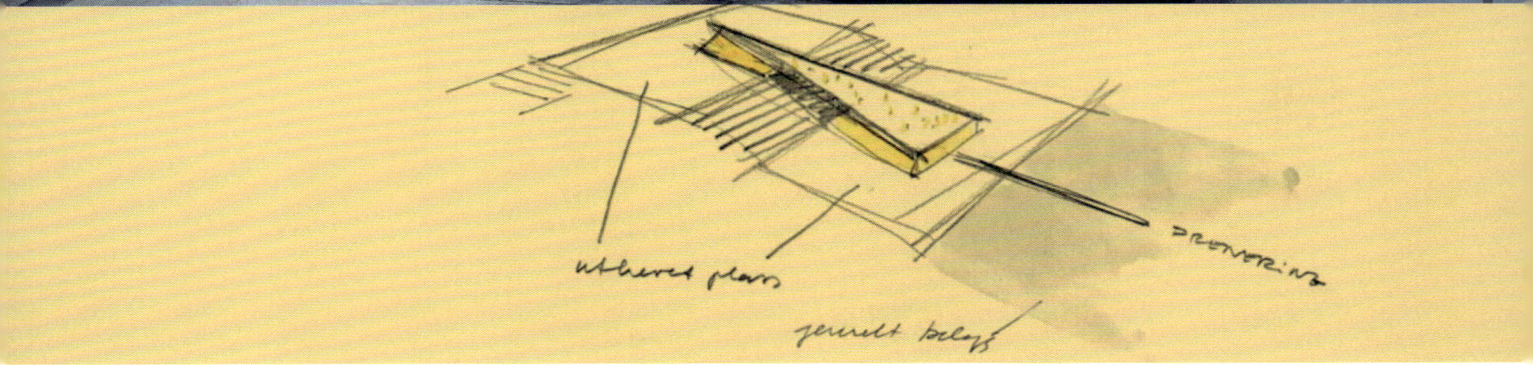

Large custom designed planters in polished gneiss is placed all along the street for about 500 meters.

They contain perennials as well as trees.

In the first quarter of the street there is a fountain connected with a large planter by a polished wall of black granite. The wall is divided by glass boxes that are lighted in the night.

OALSGATA

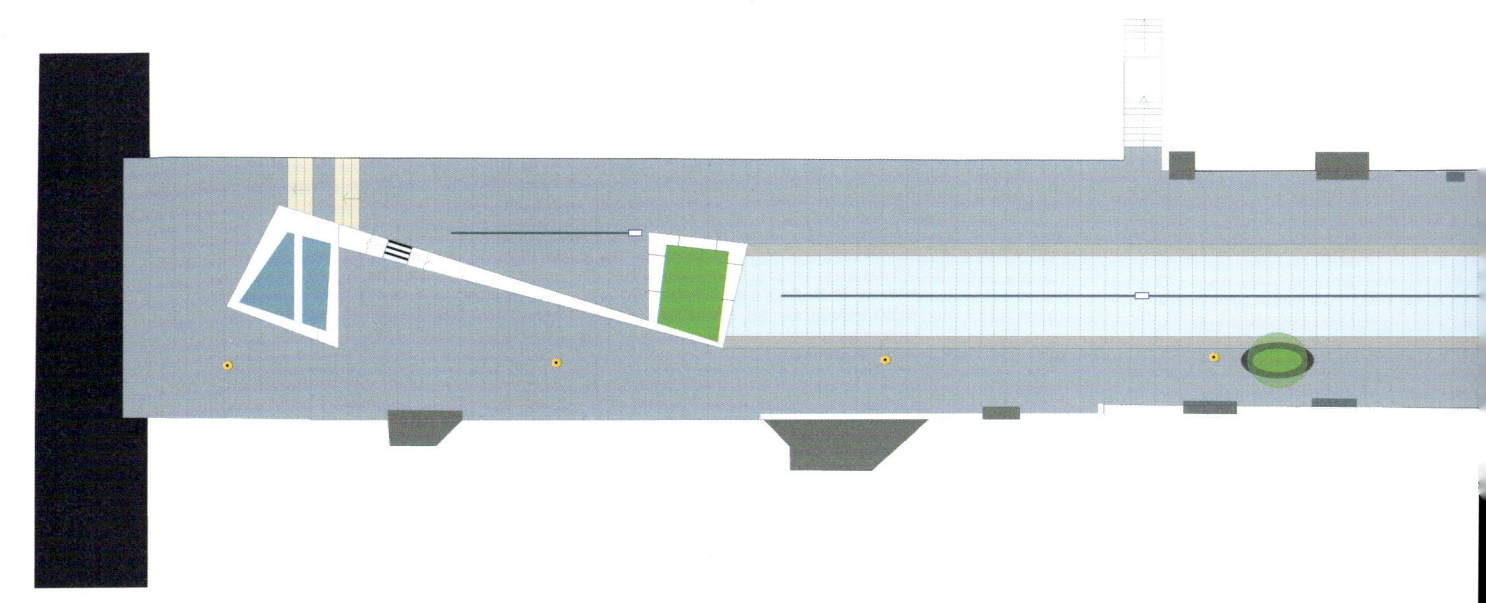

Solavegen

DNB

Minibank
DNB
SKORINGEN
AKADEMIET

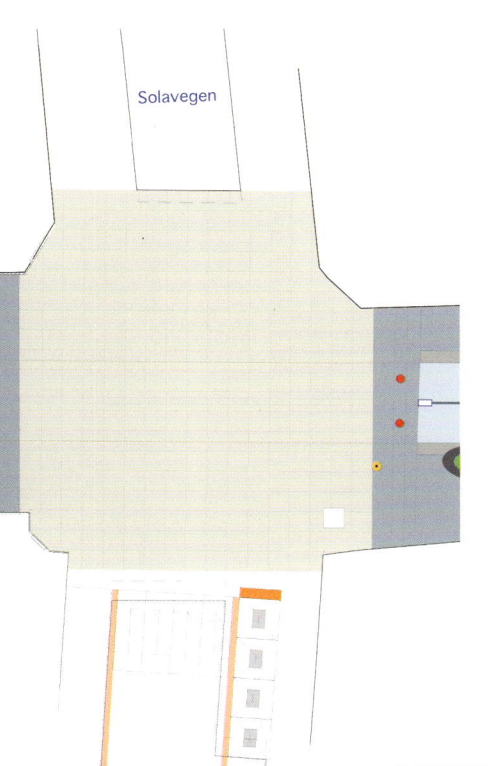

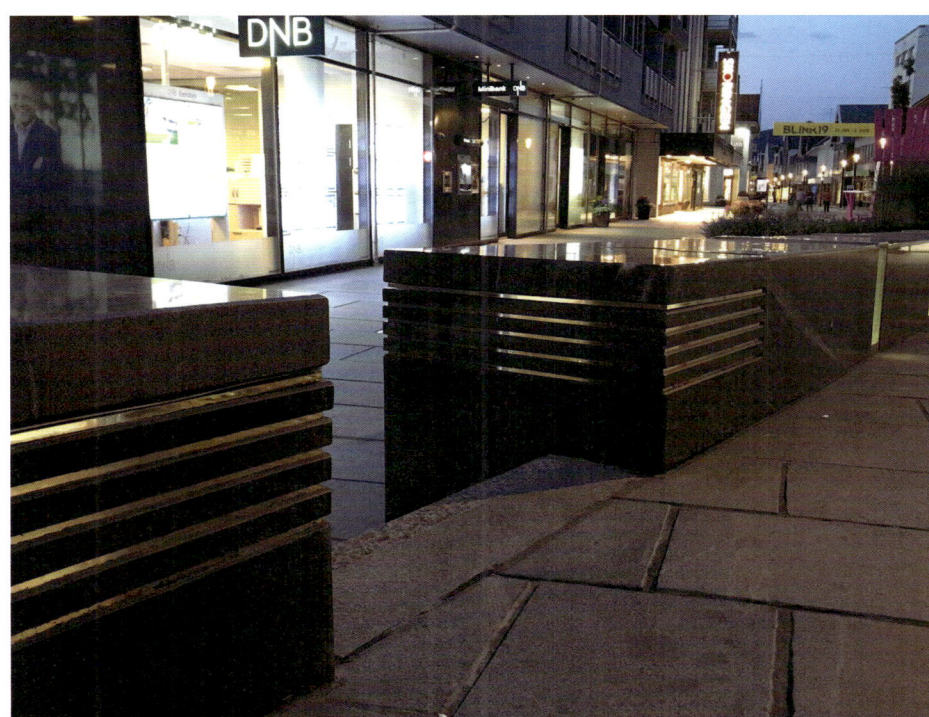

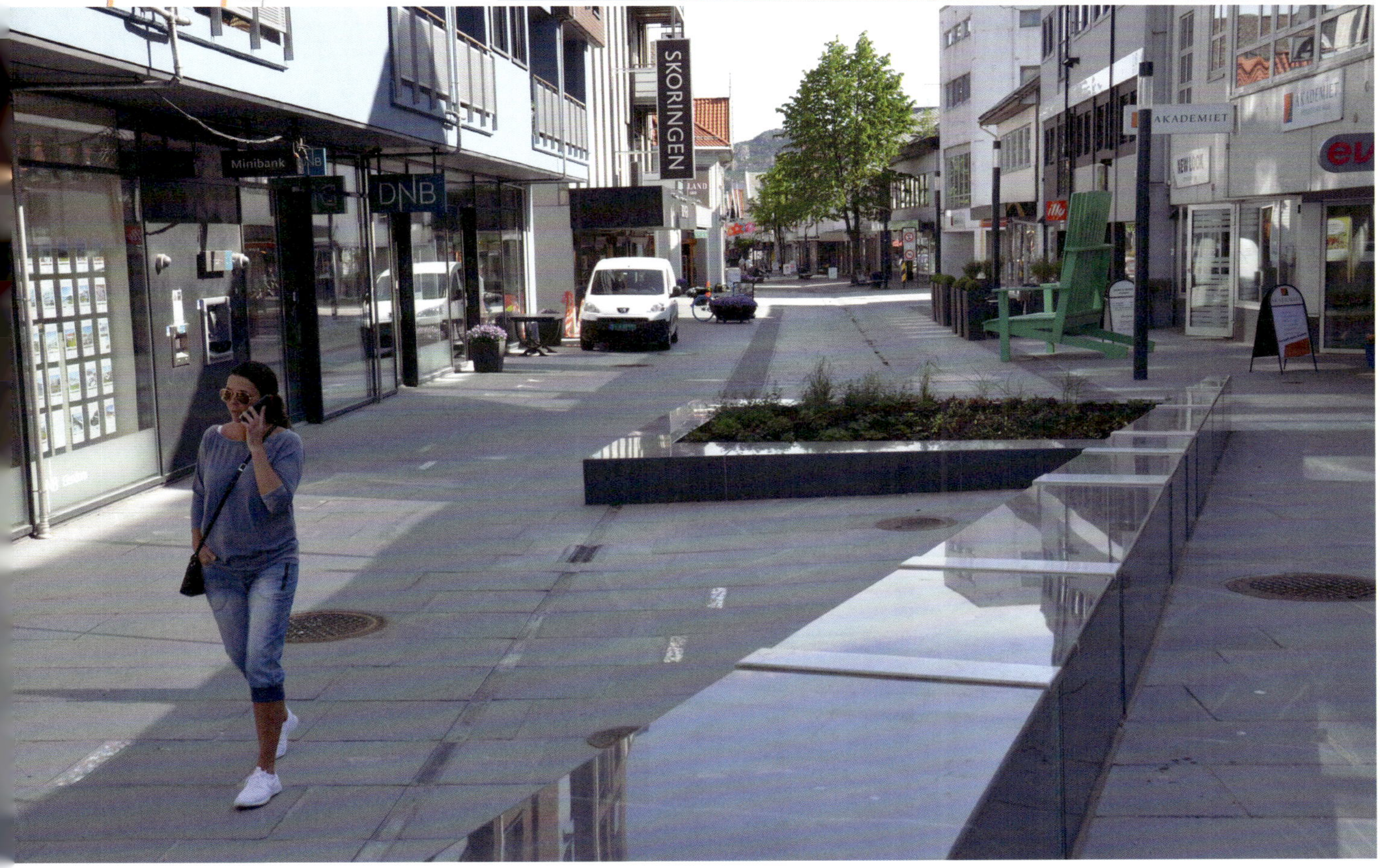

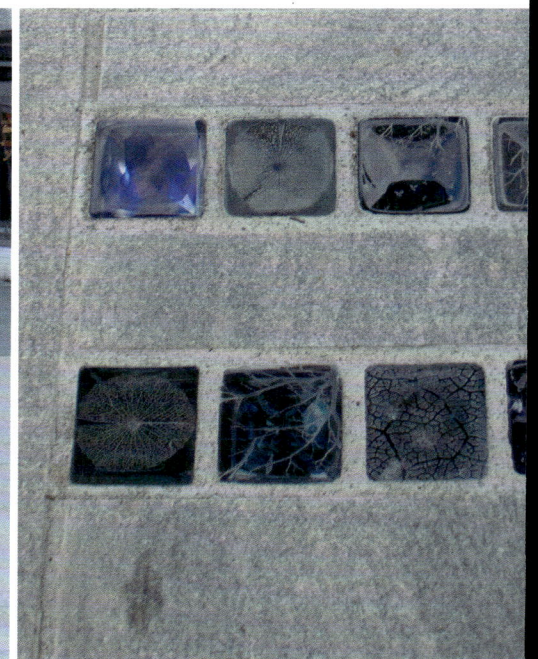

NØRRE SQUARE

VEJLE, DENMARK

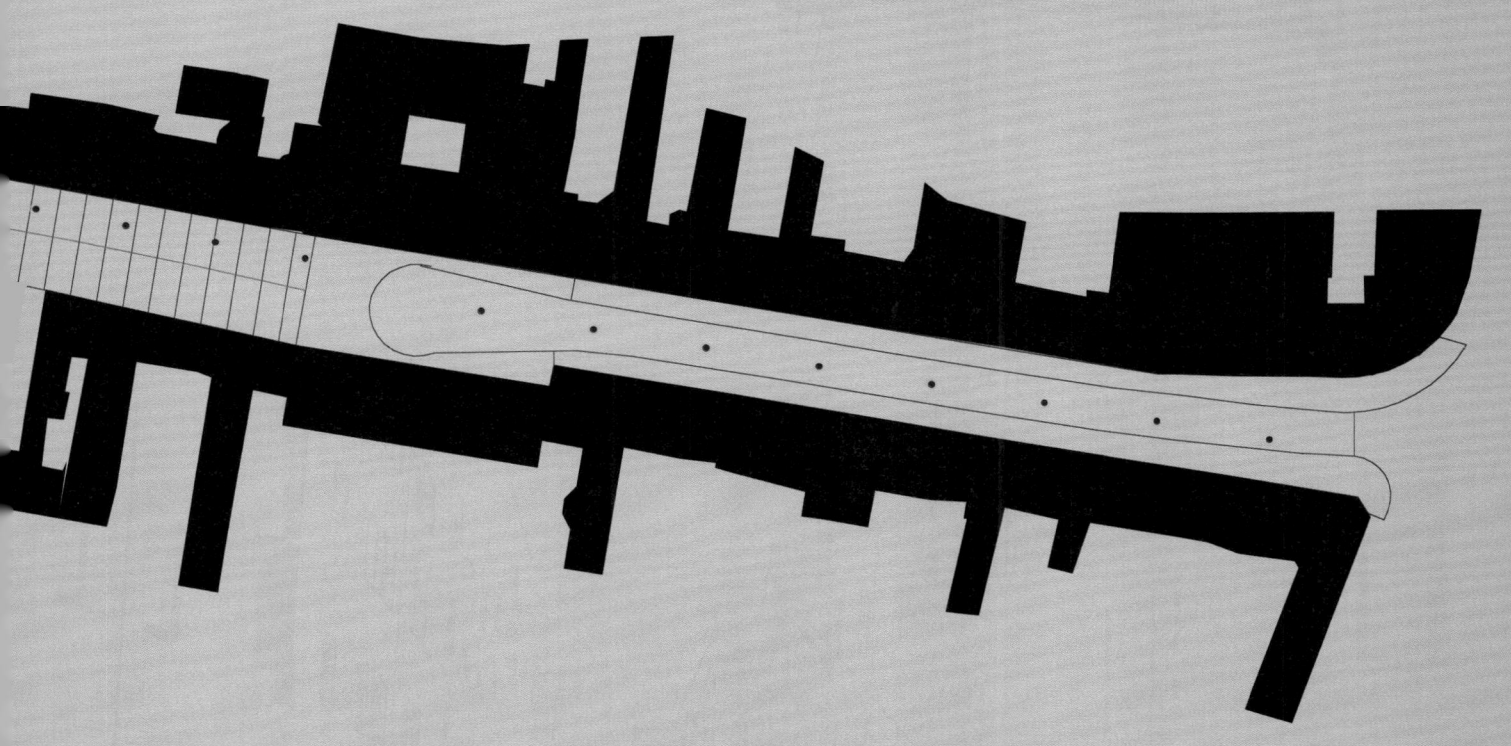

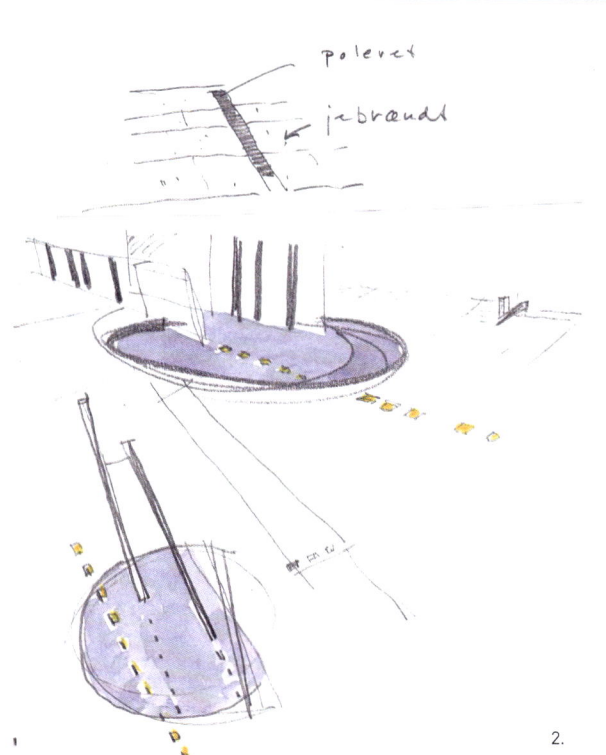

1.
2.
3.

1. Sewage bassin below the square allows only a small forest.
2. Preliminary sketch of pool.
3. Custom designed drains in brass and bronze.
4. Sketch of large granite walls covering the sewage-station.
5. Close up of the granite wall.

NØRRE SQUARE is an urban space within the medieval centre structure of Vejle. It origins from the 1400 century, and today it is connected to the main traffic route in the city. This square was to be redesigned and the old river, formerly in pipeline, to be reopened.

The square is one of the few main bus stops in the city centre, and to separate the bus-terminal from the rest of the square, three huge granite blocks are laid in a rhythmic pattern as a protective wall against the traffic. On the calmer inside, a pool, consisting of two circles, with water-jets and bronze-details, made by the sculptor Morten Stræde.

A sculptural folie has been erected in polished granite and bronze.

It covers a 1000 m^3 storm-water basin under the square.

The conceptual idea is to connect all vertical elements on the square to a centrally located curve made by an enormous bus-platform as the edge of the pedestrian area. It lies like a wave-breaker against the traffic. The square s most dominant item, is a tilted lighting pole with several flood-lights, reflectors and a stainless steel base. Flood-lights with reflectors provides the general lighting, while beamers are directed toward specific details on the square. This mast is located on a central spot where it can be observed from a distance from all adjoining streets.

Other vertical elements on the square are three large shapes of granite which are placed rythmically across the grey, marble-like plaza-floor.

They start by the lighting pole and stretches across the whole square's length and at the end you can see a bus-stop which really is a sculptural building containing pumps etc. for the Storm-water basin.

The drainage-system is polished granite drains with bronze grilles on the square.

From the buss-street crossing the square, trees are "walking" into the square. They are linden-trees with specially designed cast iron grilles.

4.

5.

1.

2.

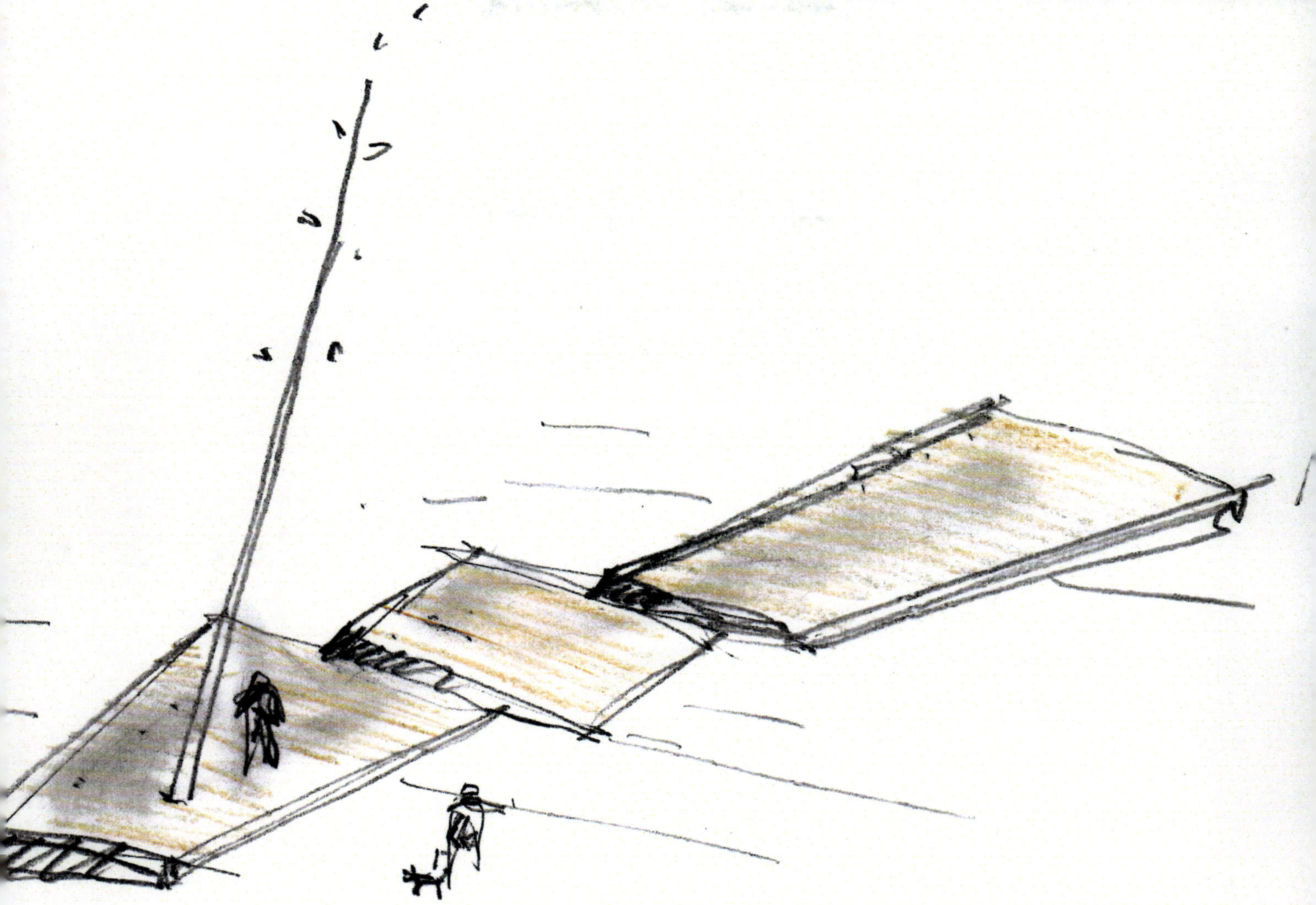

3.

1. Large granite blocks divide the calm inner square from the busy traffic.

2. / 3. Sketches of the granite blocks.

BROSTEN I KJØREBANE

BROSTEN OG GRANITHELLER LANGS OMLØBSÅEN

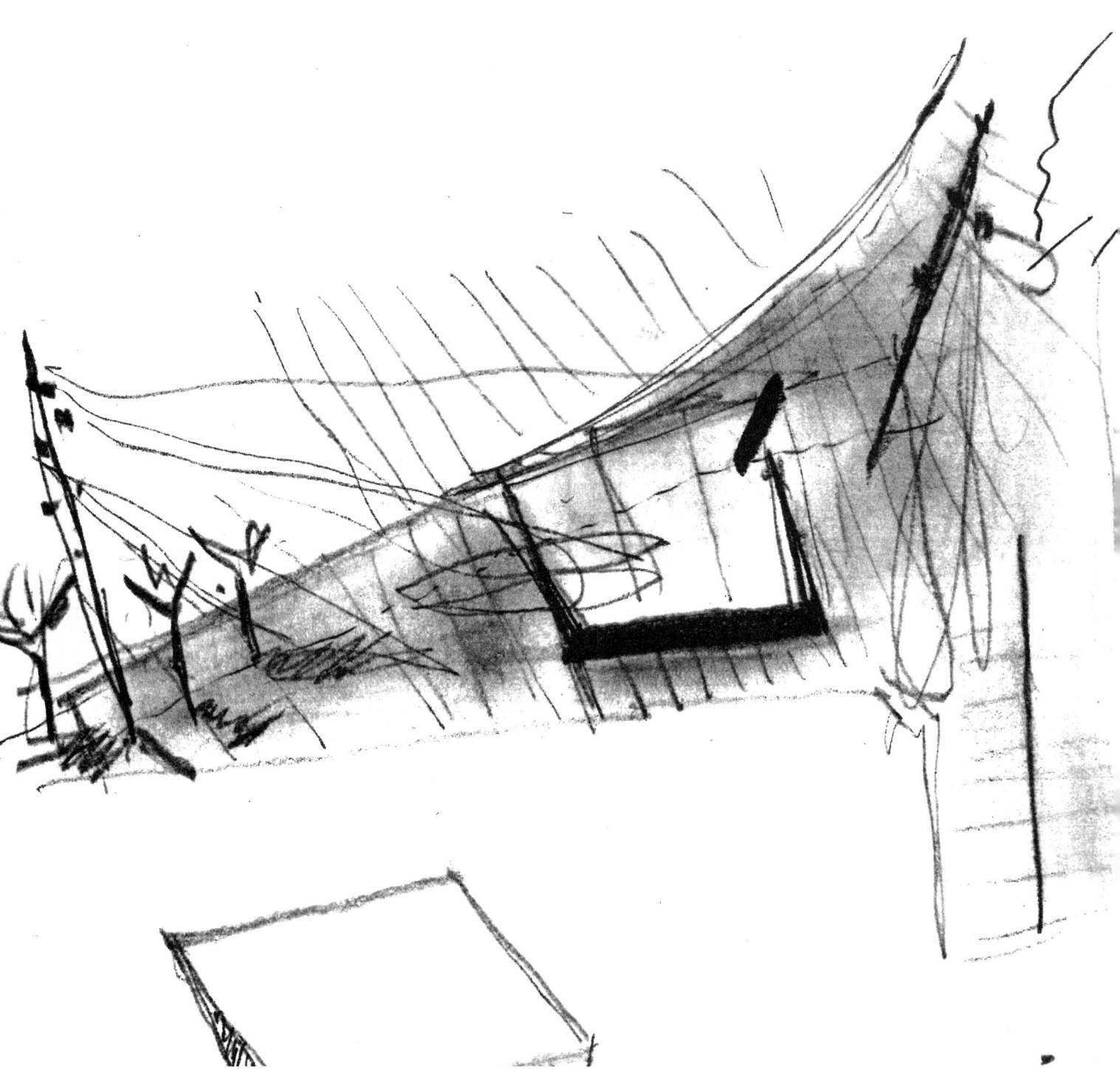

Preliminary sketch of the square.

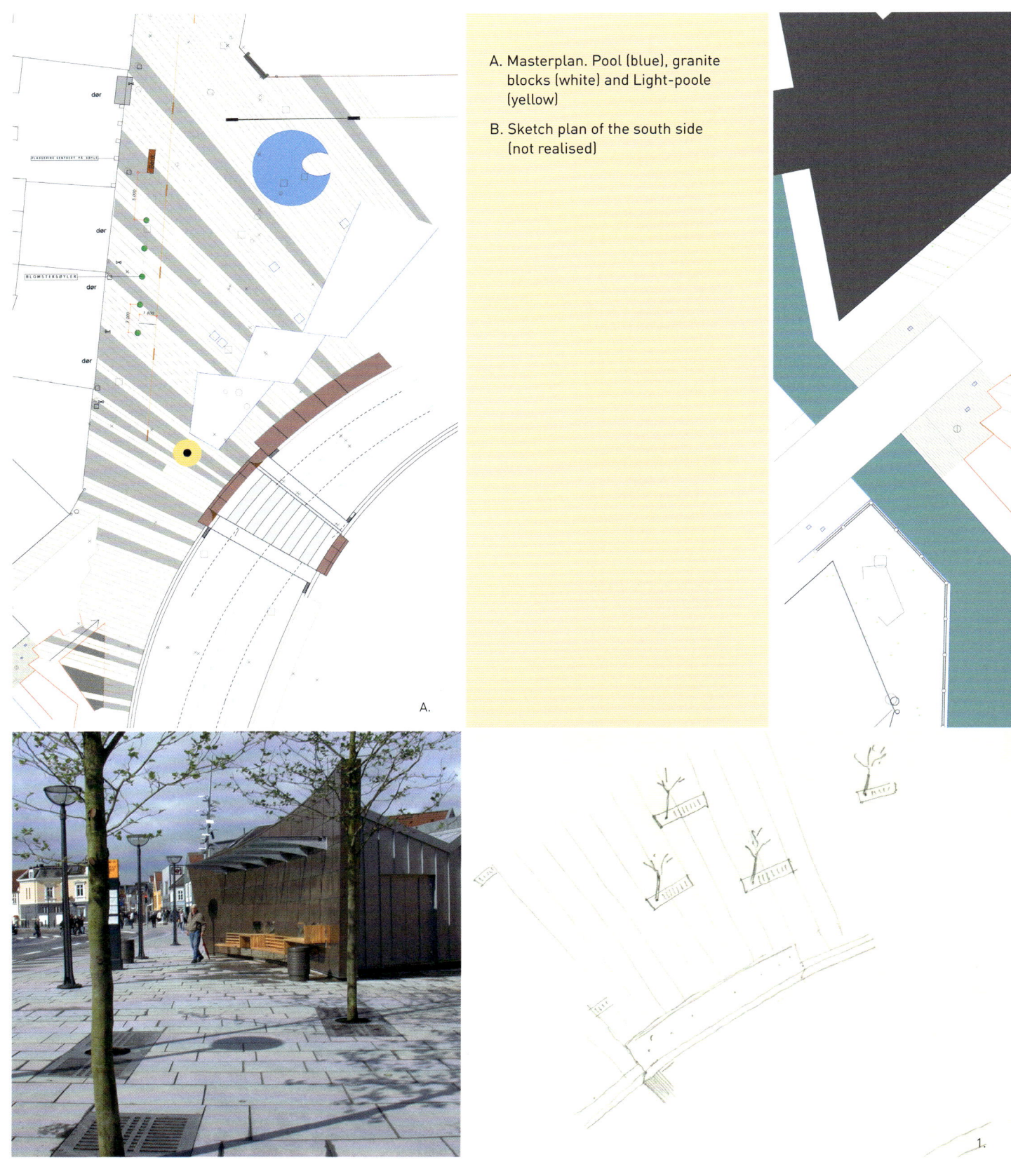

A. Masterplan. Pool (blue), granite blocks (white) and Light-poole (yellow)

B. Sketch plan of the south side (not realised)

1. The sewage-station also functions as a bus-stop.
2. Sketch of tree-grilles, later casted in iron.

ORLA LEHMANN STREET

VEJLE, DENMARK

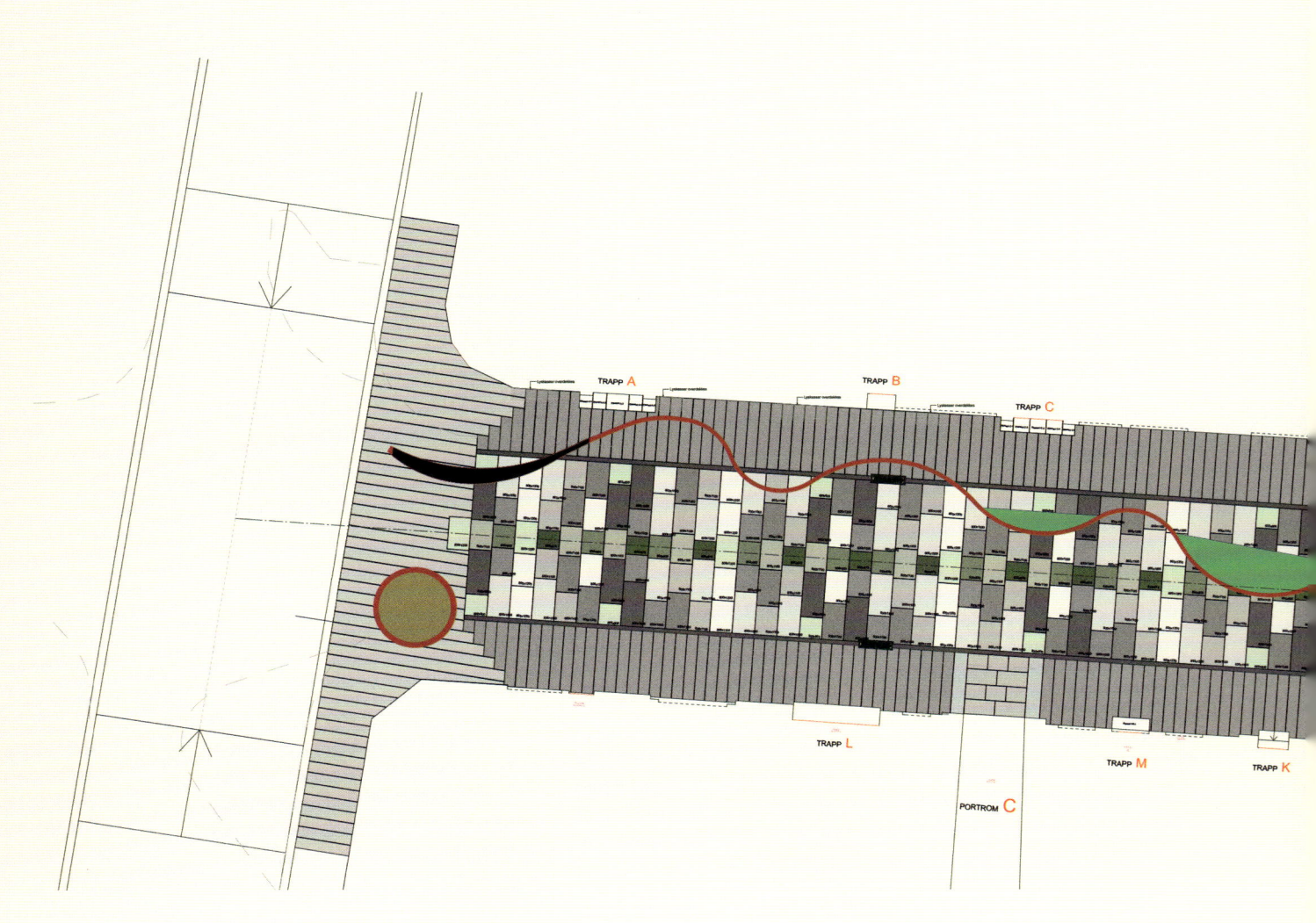

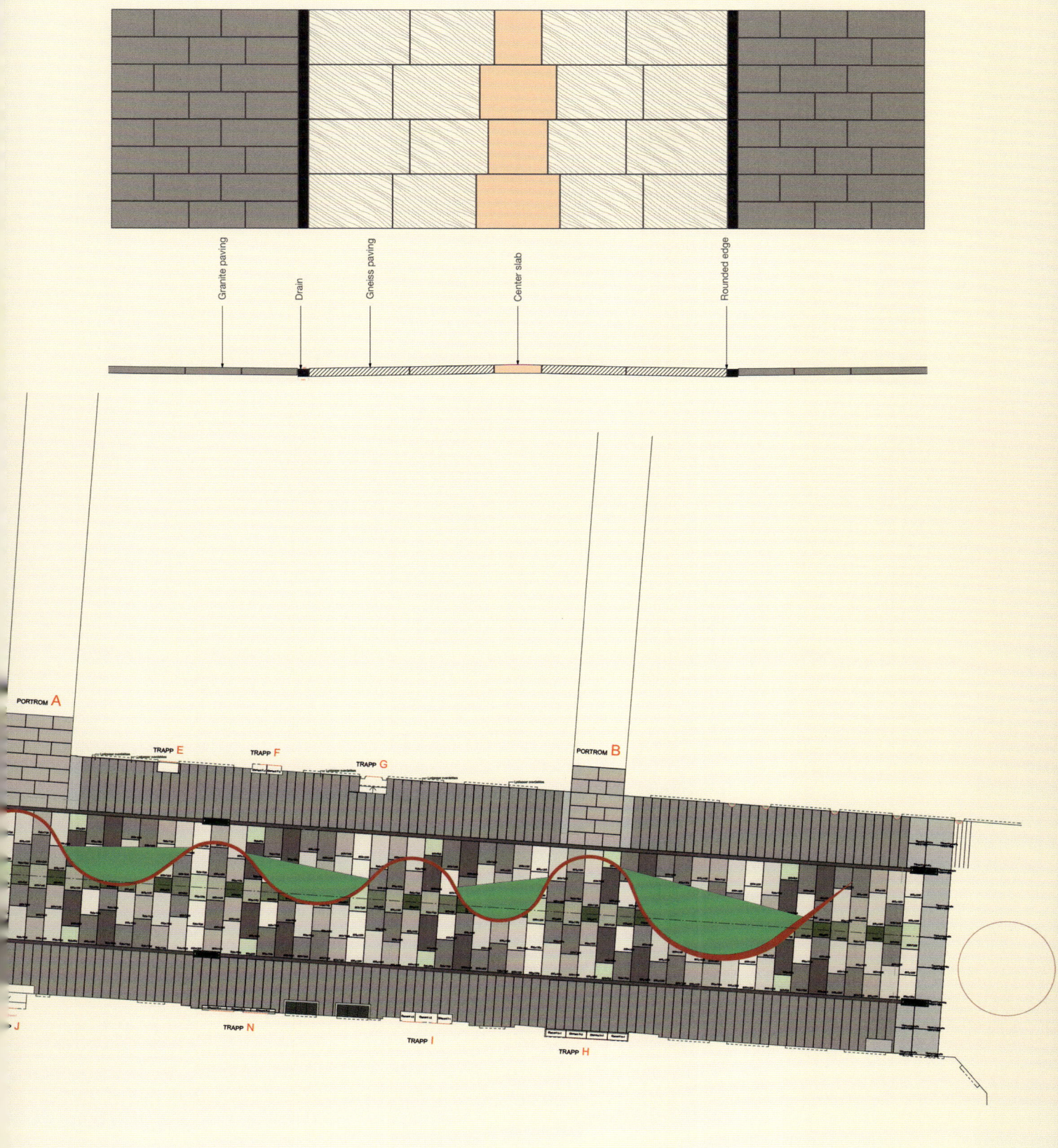

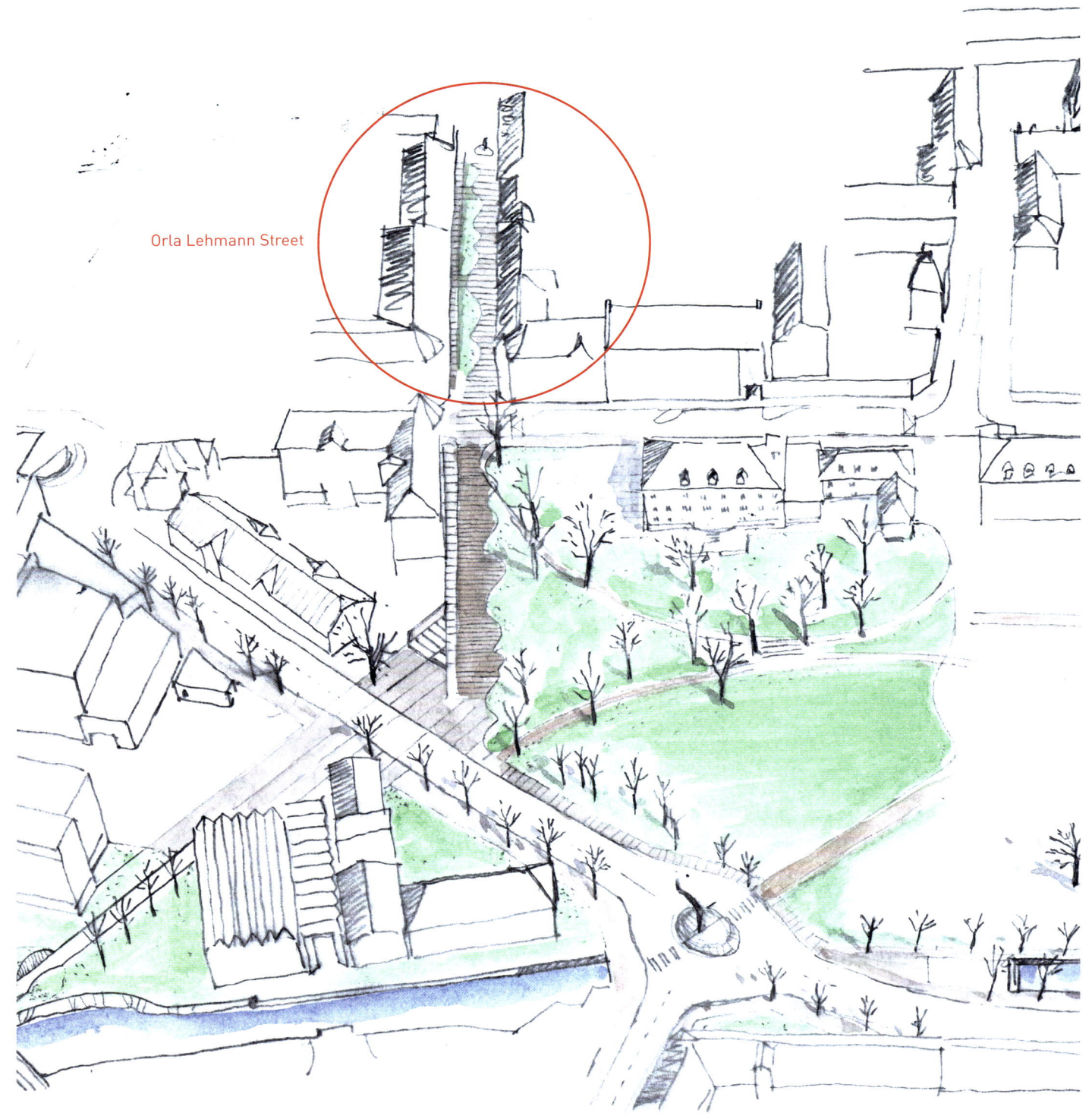

Sketch showing the connection between existing park and the street

Conceptual idea shown on drone image.

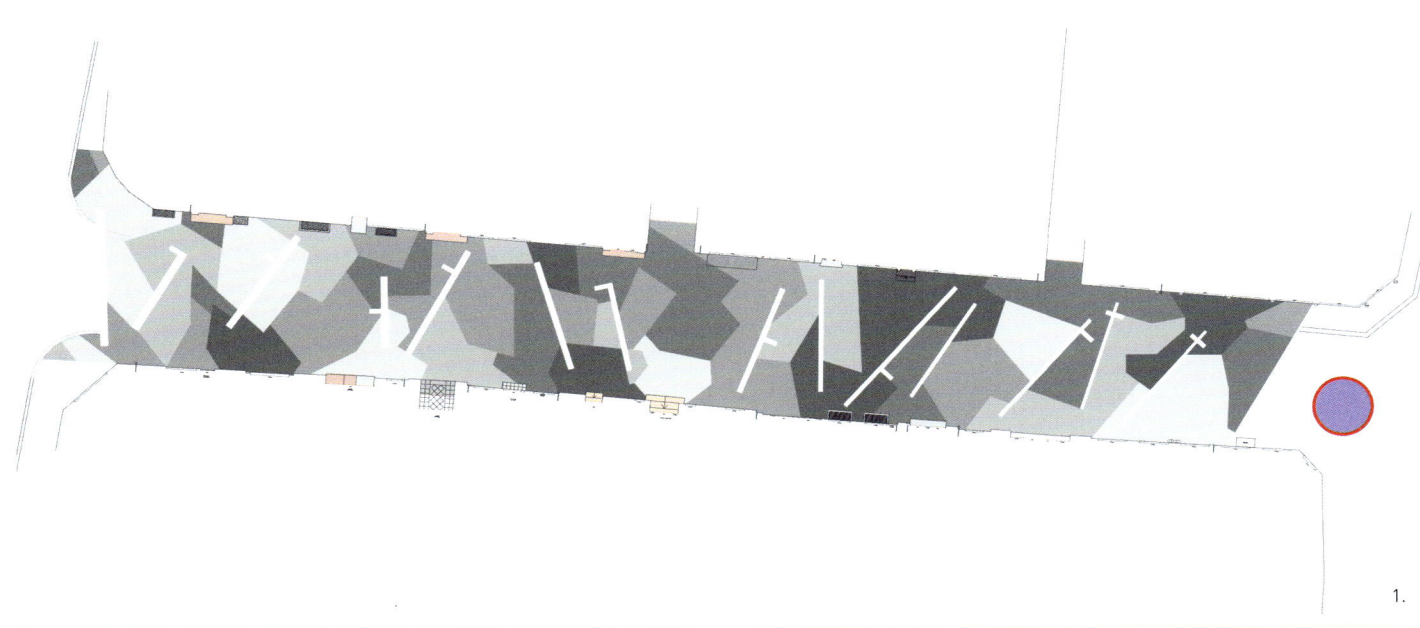

1. Early paving suggestion.
2. Sketch plan.
3. Final plan.
4. After intervention.

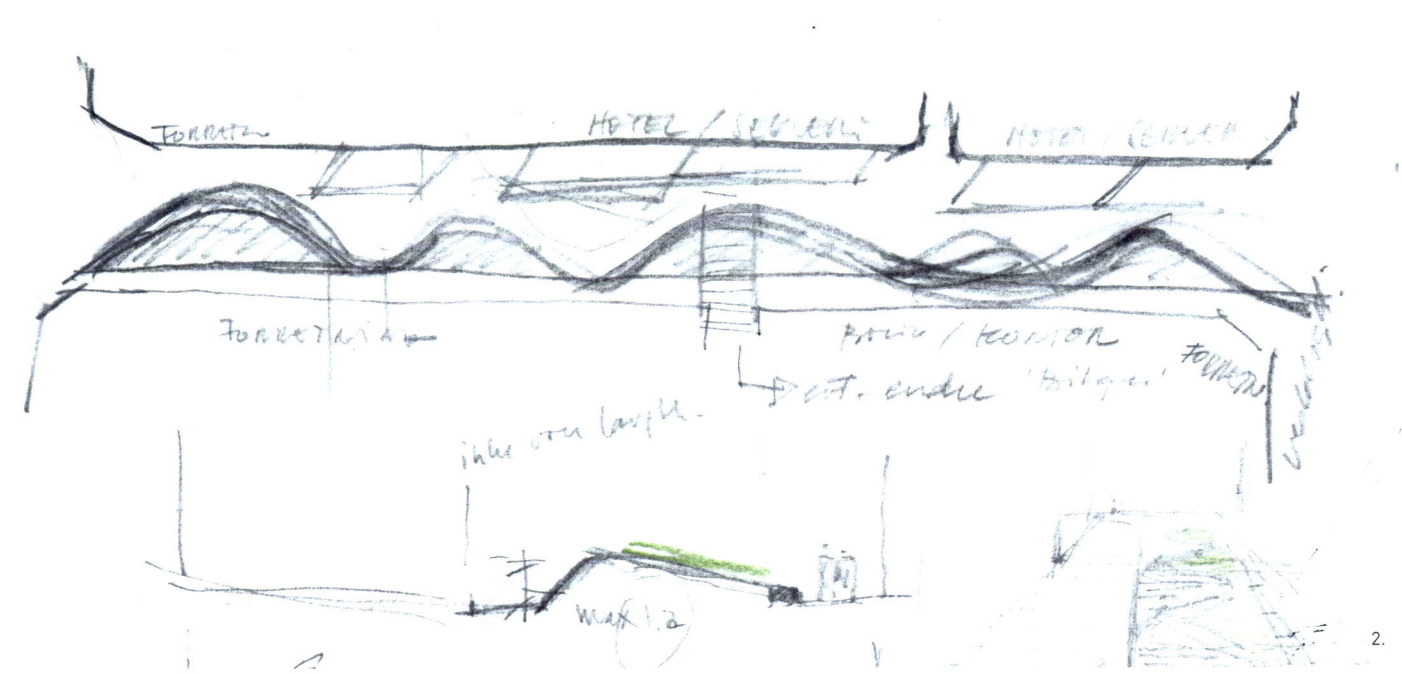

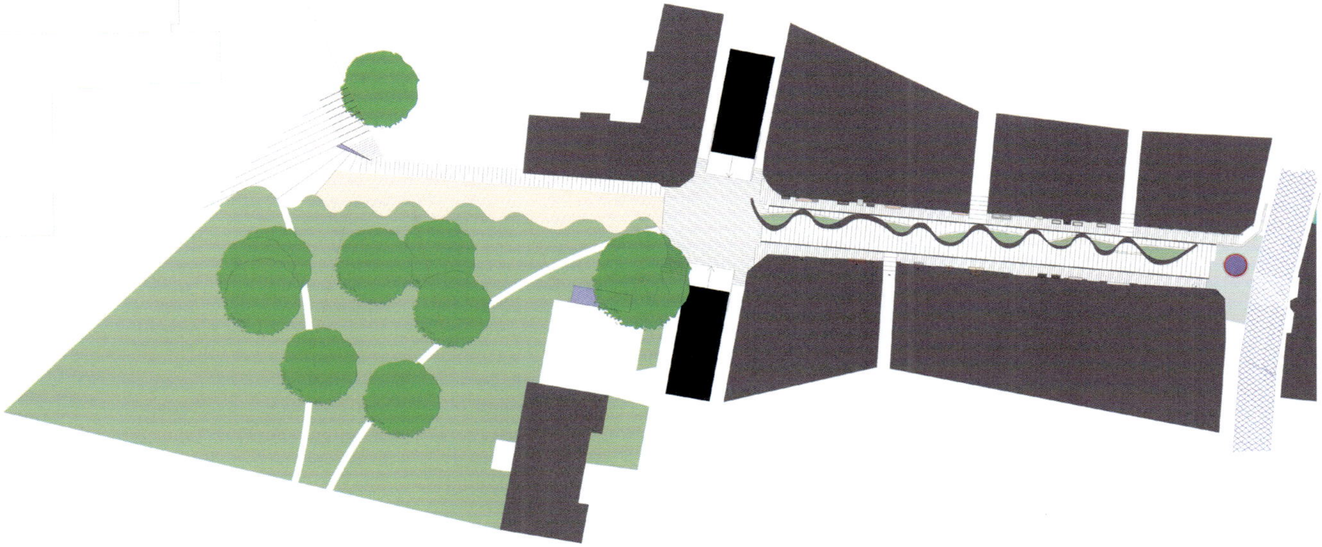

151

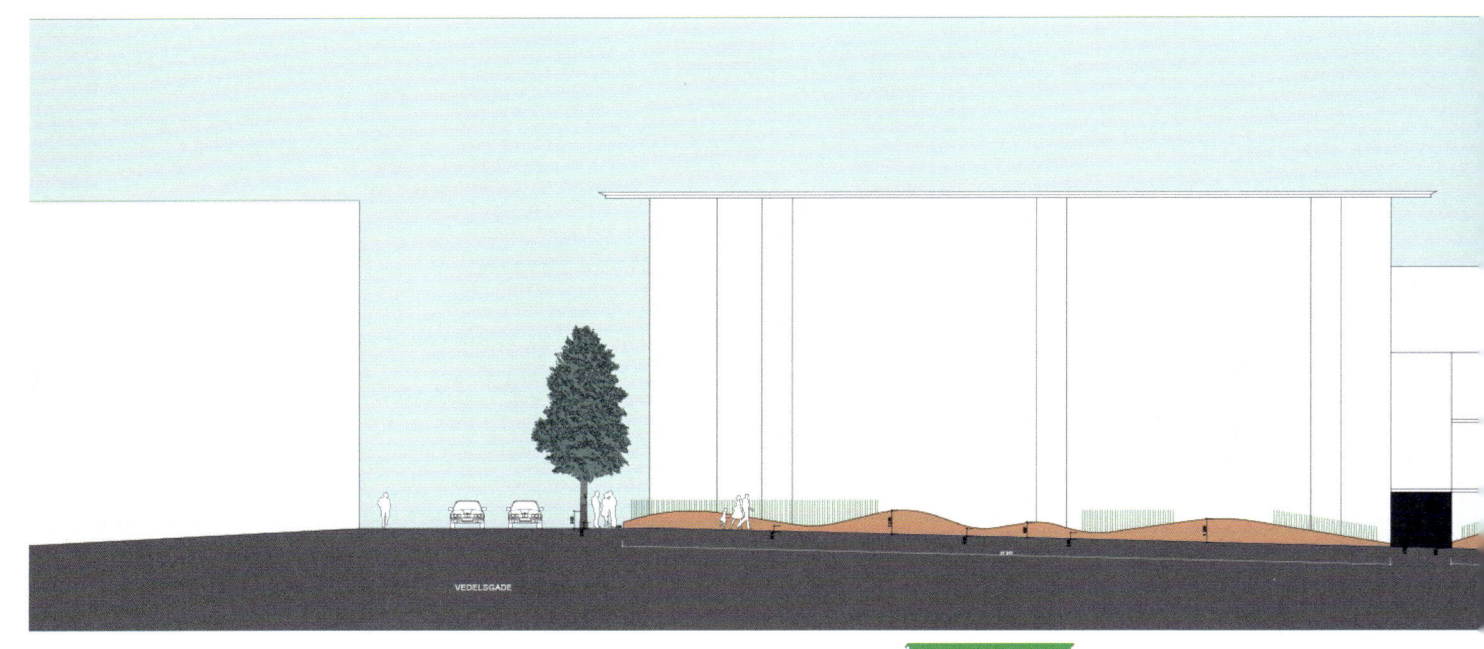

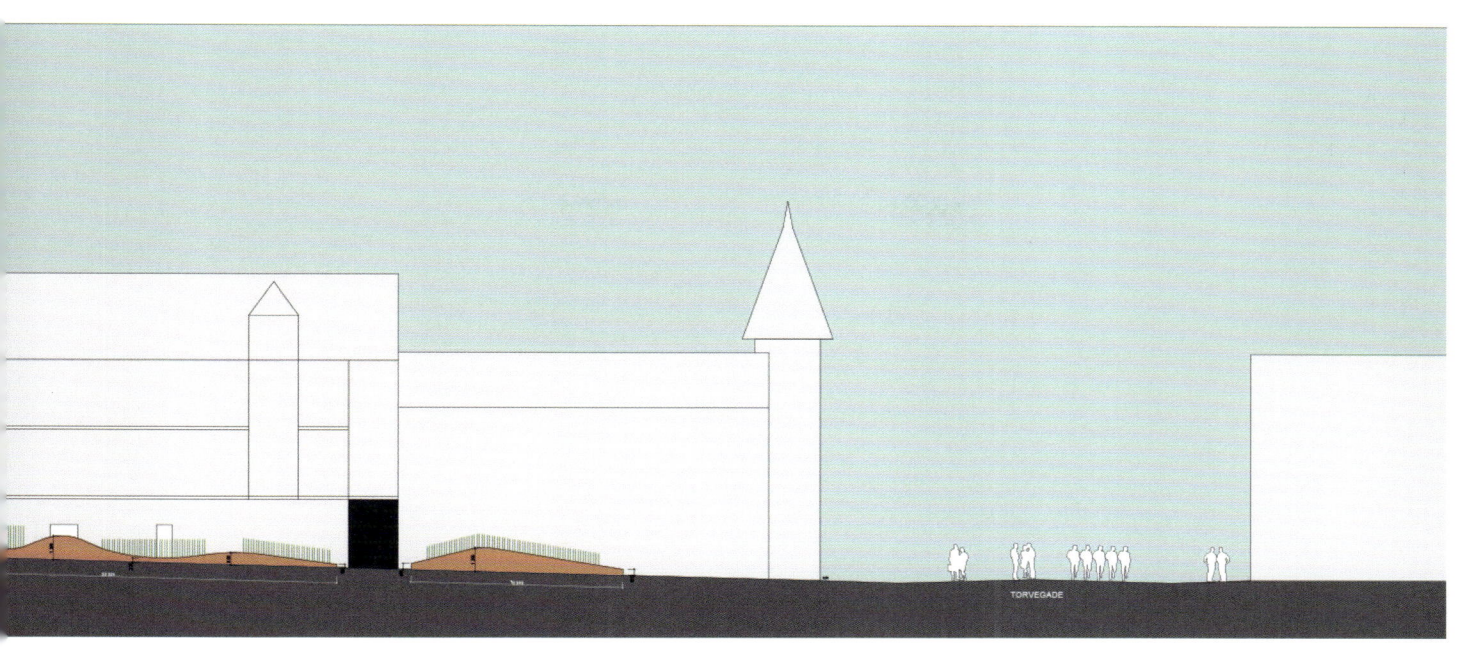

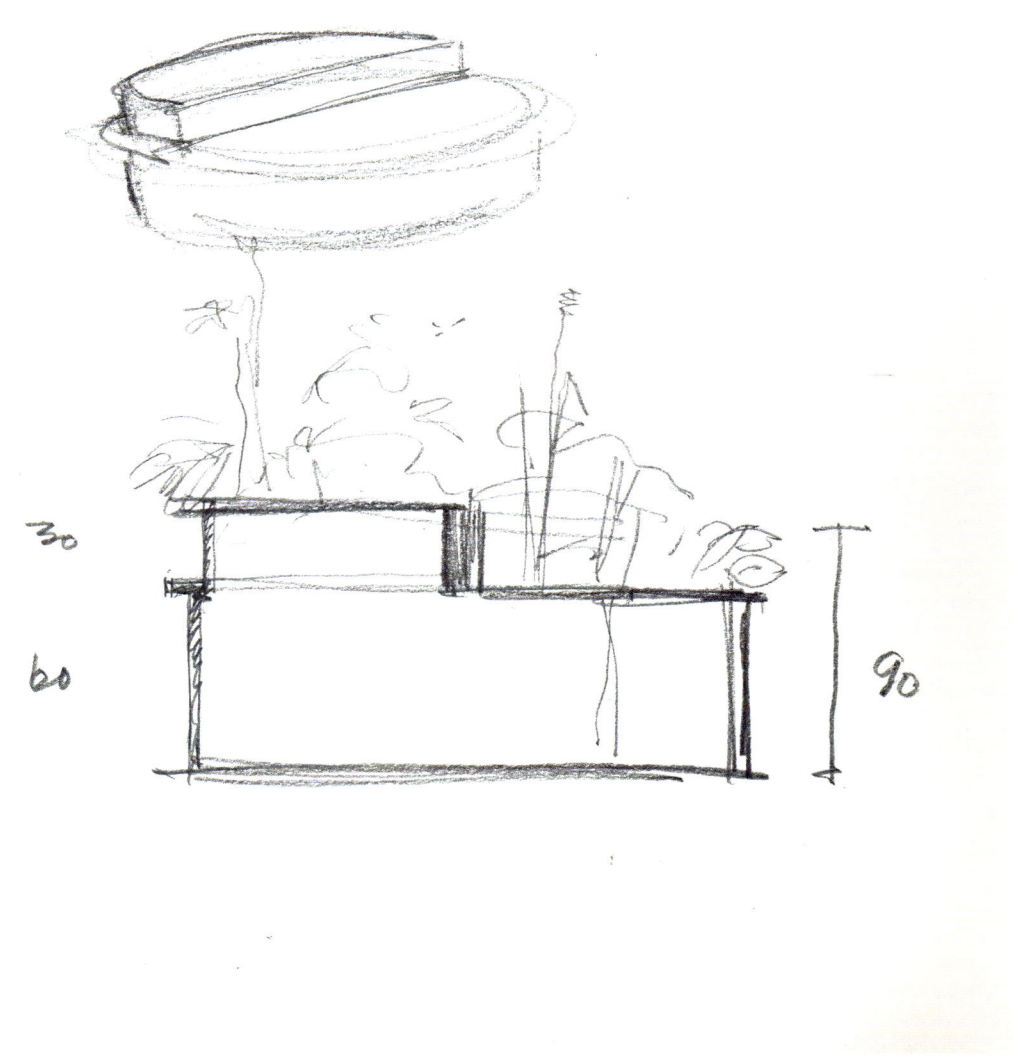

Early sketch of bronze planter.

Sketch of streetscape with bronze planter.

After intervention.

ORLA LEHMANN STREET lies in the medieval centre of the city of Vejle, Denmark.

It is narrow and face west, so in the afternoon, there is a ray of sun for a short while. But mostly the street feels dark.

My intention was to challenge this and make a «tropical garden». The many doors and gates was another challenge. Cars had to pass by. The solution was a snake-like wall of bronze with access to the adjacent buildings, and with palms and perennials in large planters.

The intro from west is a giant bronze planter embraced by the bronze-snake.

1. The bronze form ends integrated in the paving.
2. Detail plan.

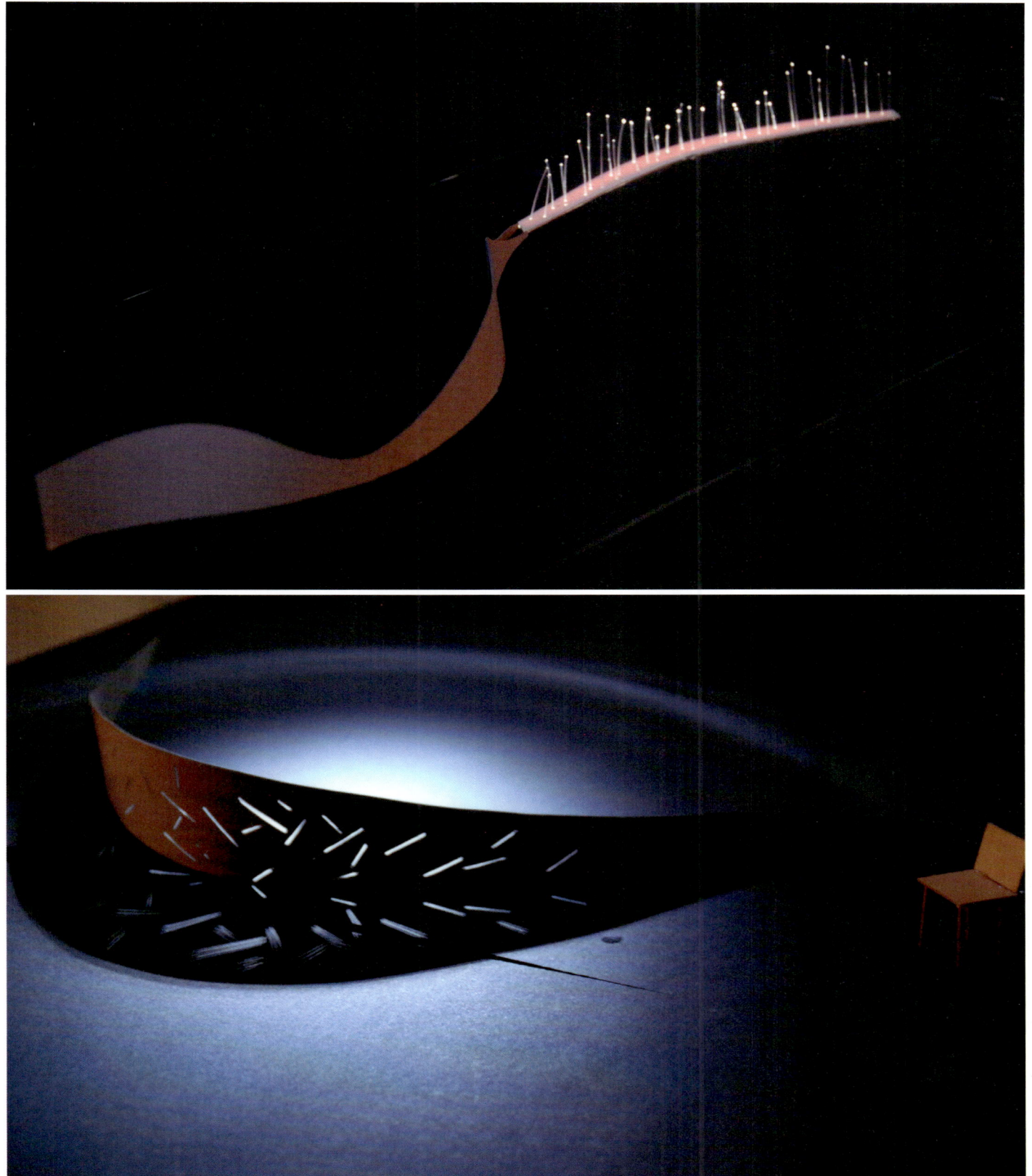

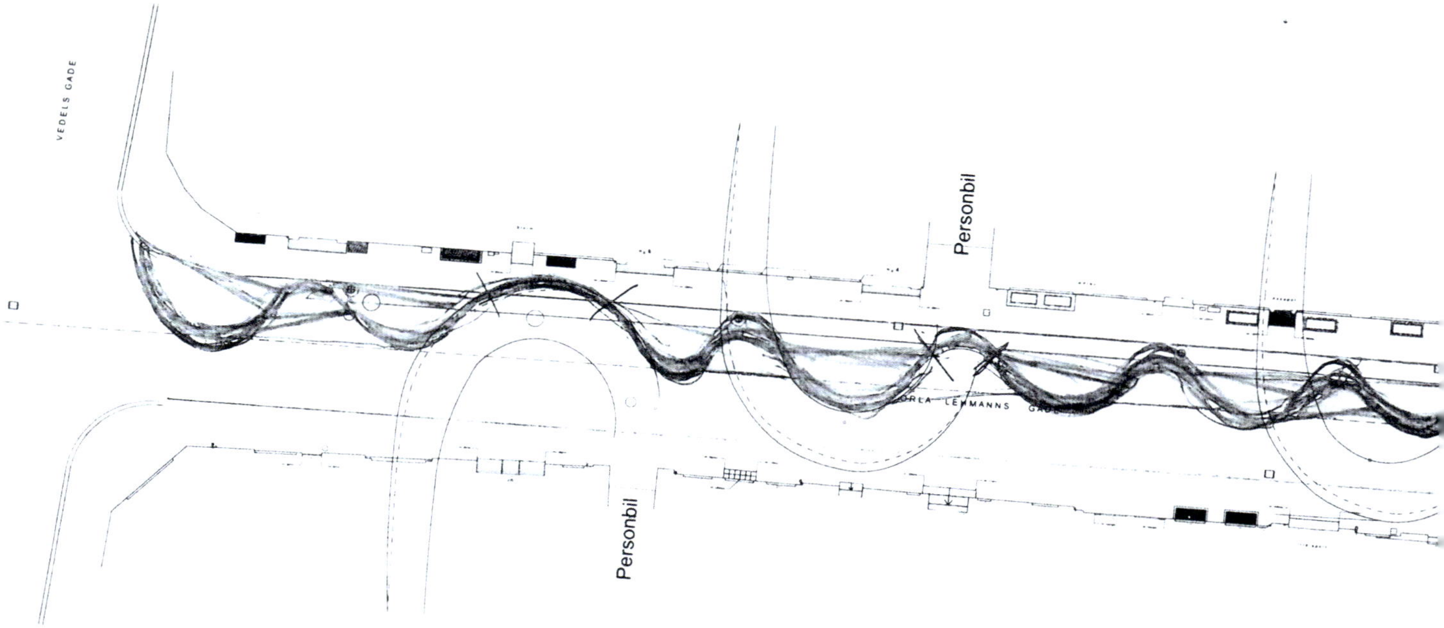

Sketch showing building entrances corresponding with the bronze form.

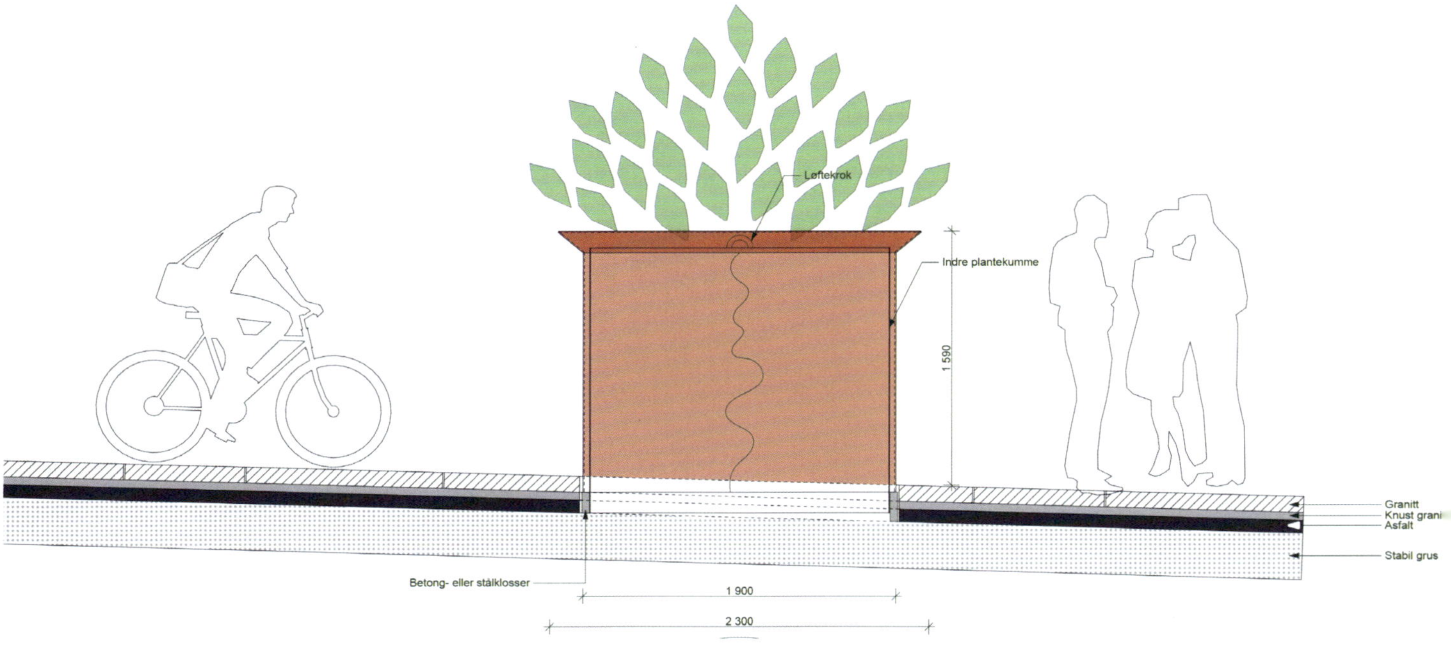

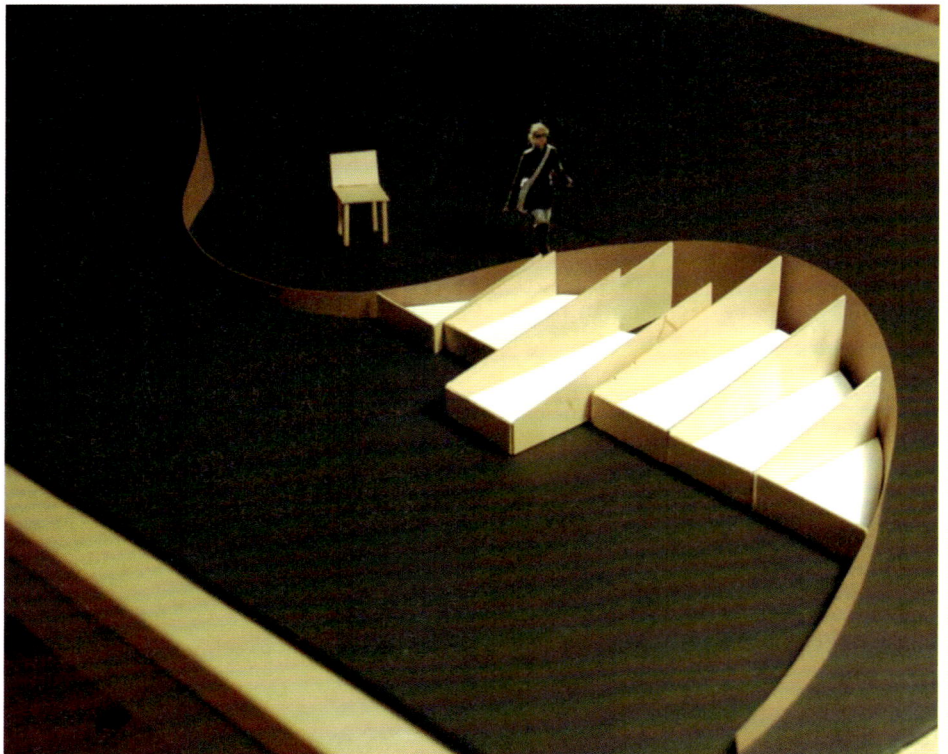

The bronze planters curves along the street.

Inside the bronze forms there are several boxes containing growth medium, weighing up to 2 metric tons.

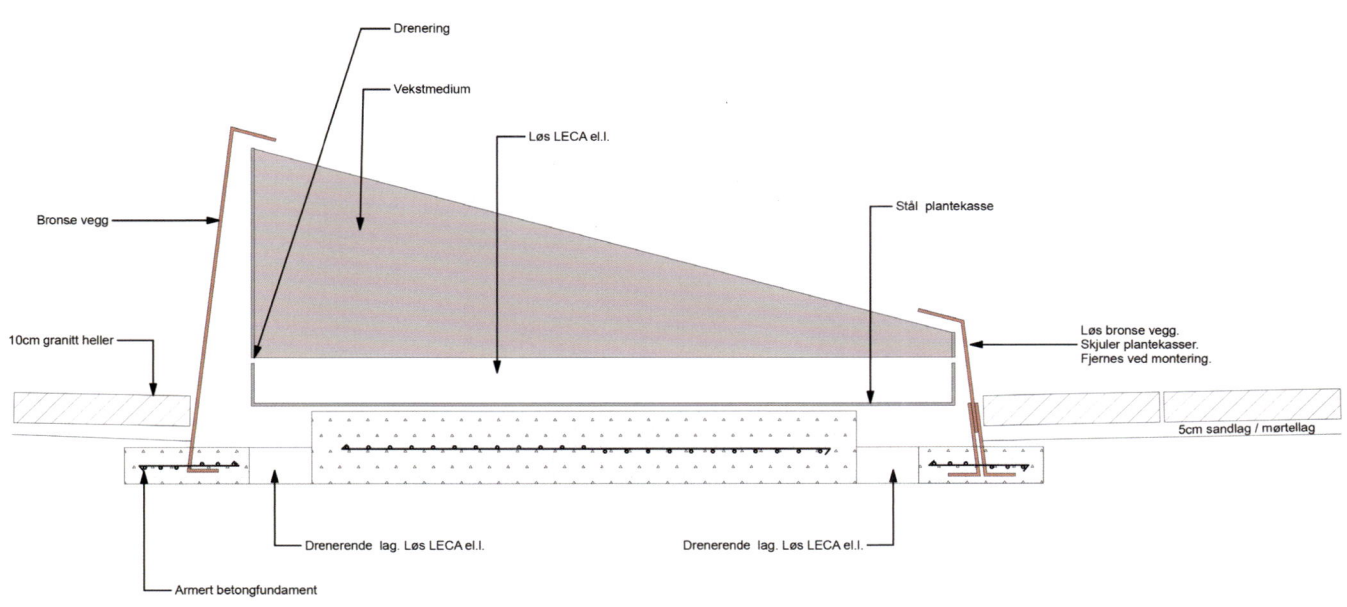

The boxes can be removed for maintenance as shown in the model.

Early proposals.

STAIRS TO THE CHURCH OF ST. JOHN

BERGEN, NORWAY

Acryl on canvas. Base for the paving pattern.

THE CHURCH OF ST.JOHN is Bergen's largest. Situated on top of West Market Street the church tower dominates and facing the city centre. Light from the buildings red brick facade reflects a warm light towards the city´s heart in the night.

The church is situated in the city's multicultural centre with the university close by. Thus a lot of pedestrians crowds the stairs.

The upper part was constructed in the 1880s. These granite steps was both too steep av too short. They had to be reconstructed, reusing the old granite curb-stones.

The lower part had ever since medieval times been a steep slope with a gradient of 25°. The task was to redesign the slope to walkable stairs to the church with entrance to buildings and gardens on both sides.

Because the entrances on the sides were on different levels, the stairs to St.John had to be a combination of normal steps for regular transport, and large ones suitable for staying.

A result was that the lower part of the site contained a serie of repos. The paving of these was constructed with five different granite-colours based on oil on-canvas sketches. As were the rest of the platforms further up hill.

A major demand from the authorities was that the new construction also should contain water-elements.

The water-system starts from a circular basin. 29 water-jets disappear in a custom designed bronze drain.

On the next, lower level, the water reappears in two outlets in granite. From there it runs through open drains and cascades until it runs down a bronze drain before it reappears in another water jet and a 2 meter high waterfall. In the end the water rests in two basins at the foot of the stairs.

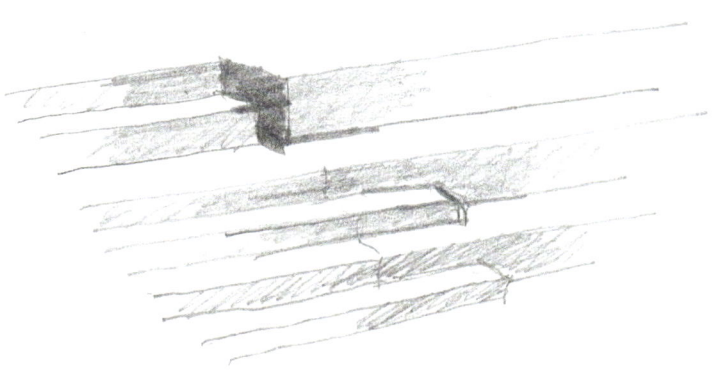

variere modellering

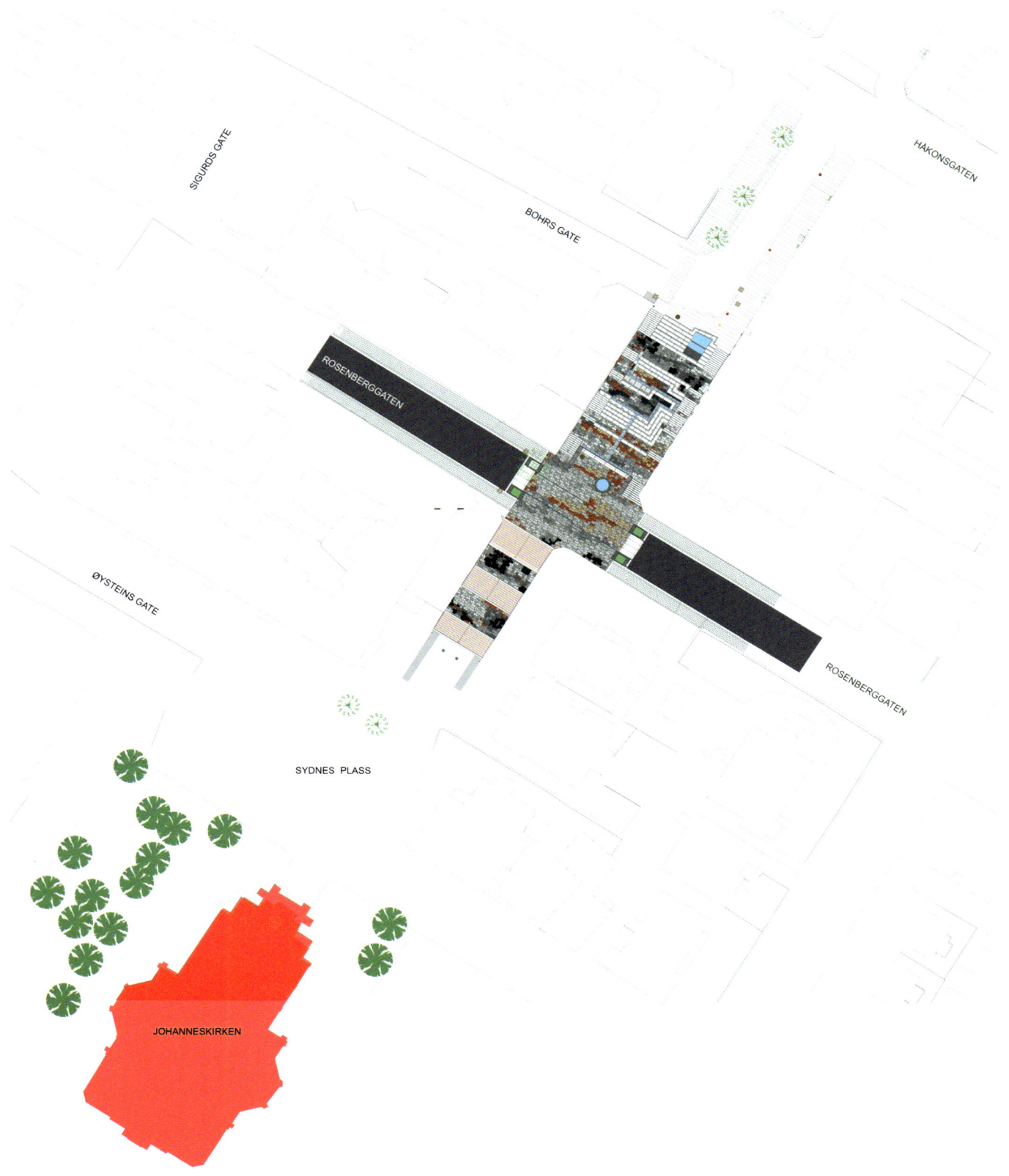

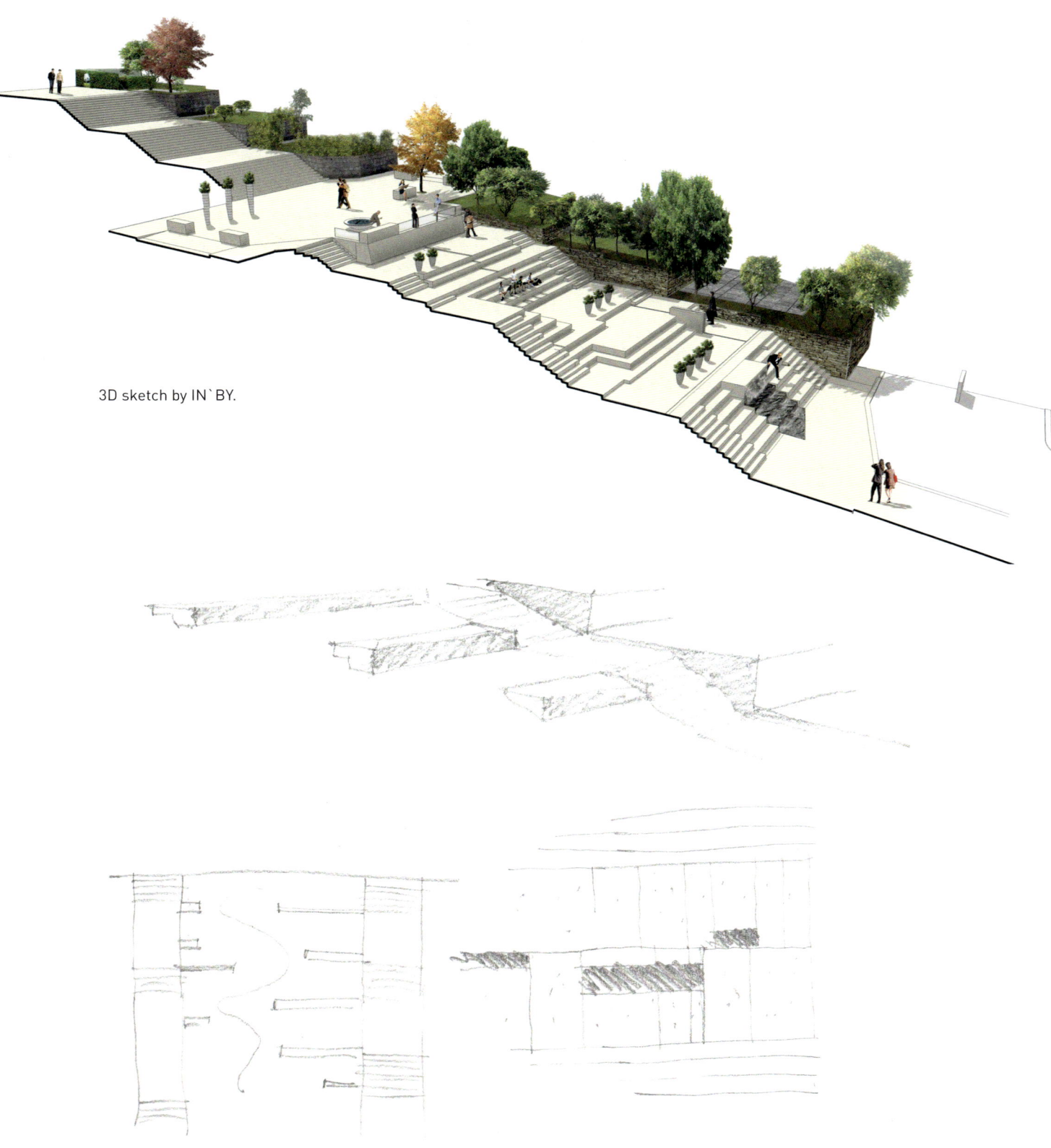

3D sketch by IN`BY.

Acryl on canvas. Base for the paving pattern.

Modules of paving pattern.

3D sketch by IN`BY.

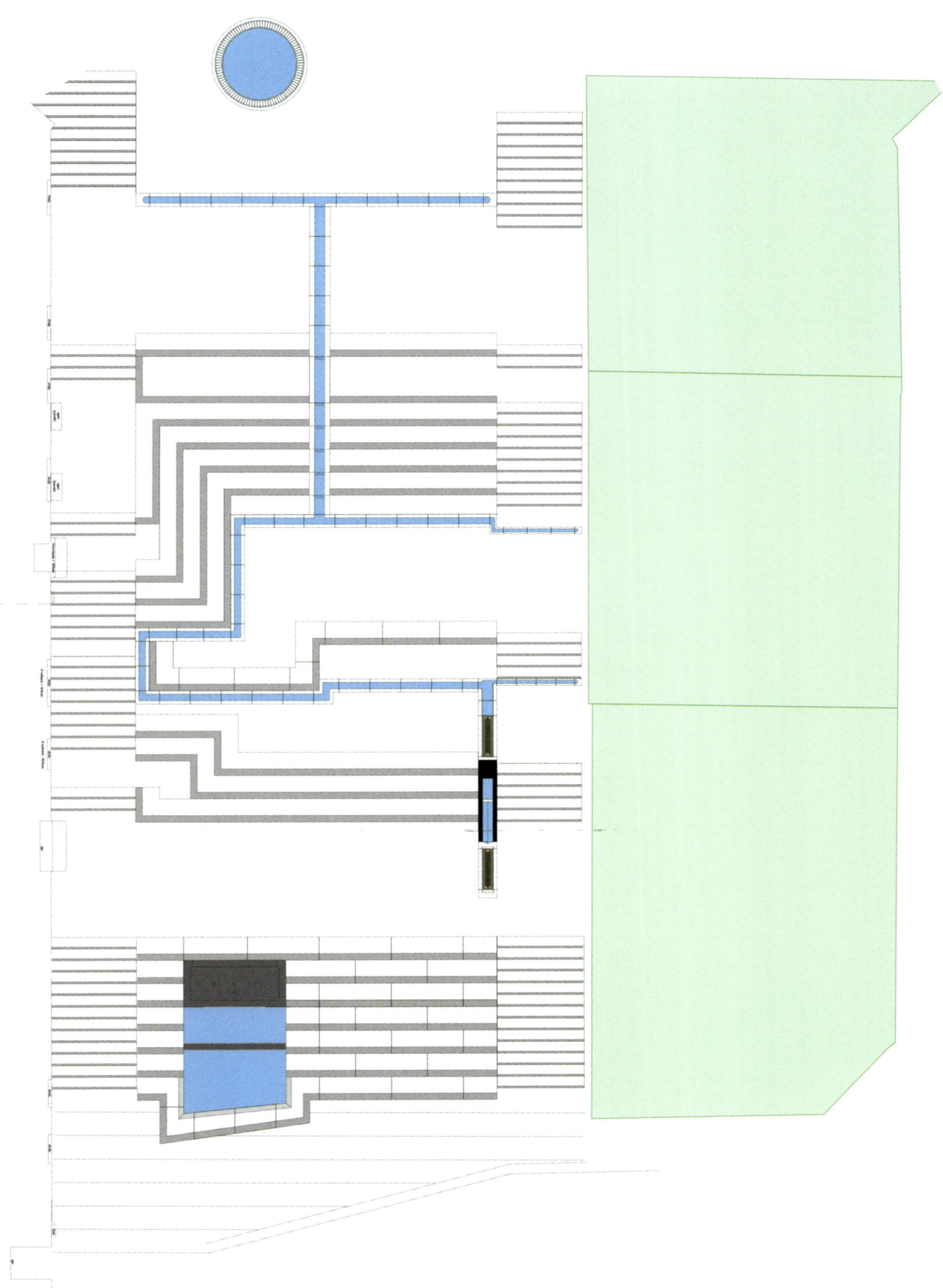

177

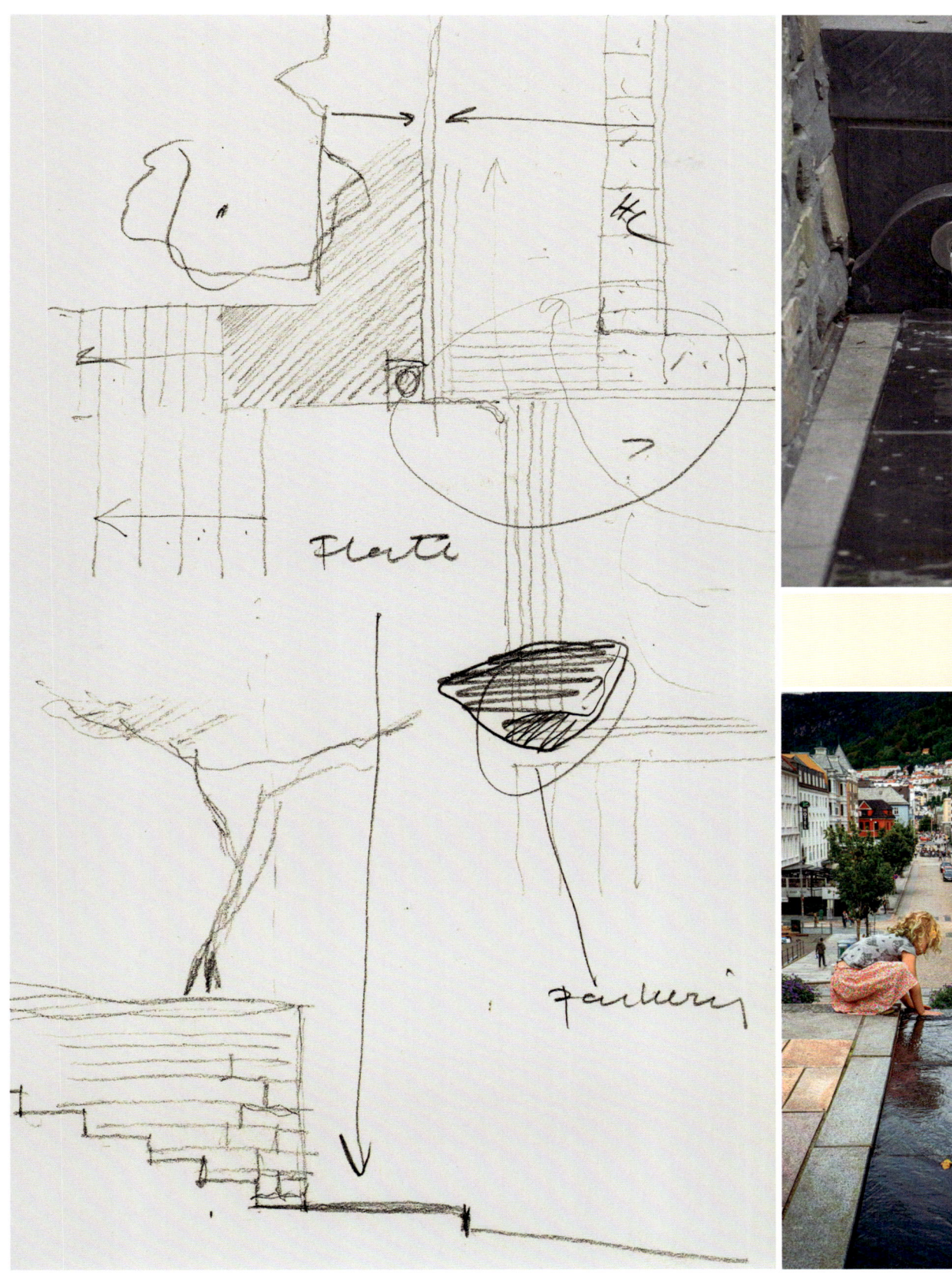

180

181

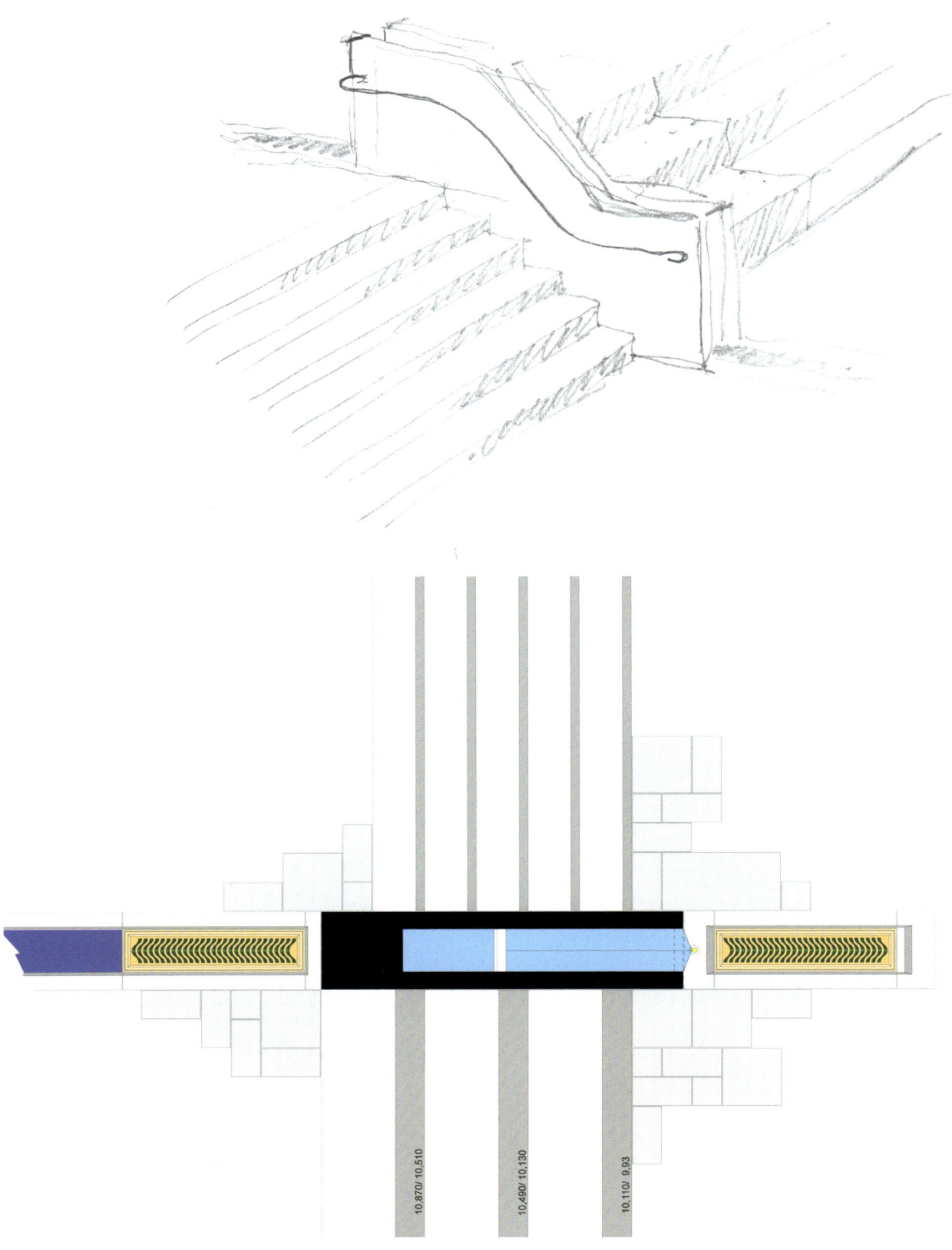

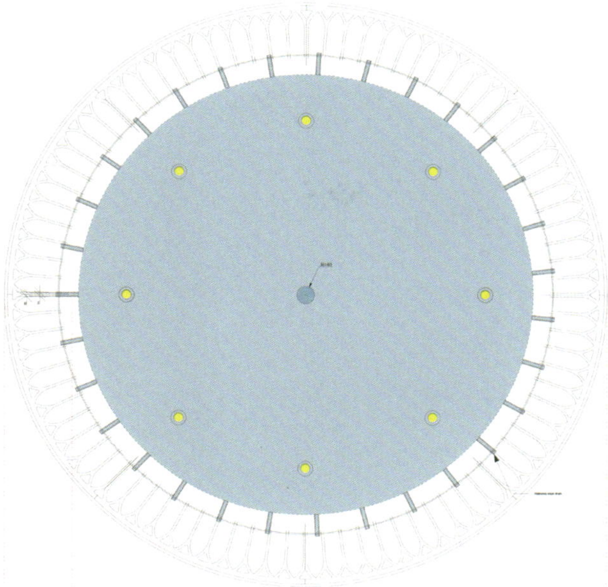

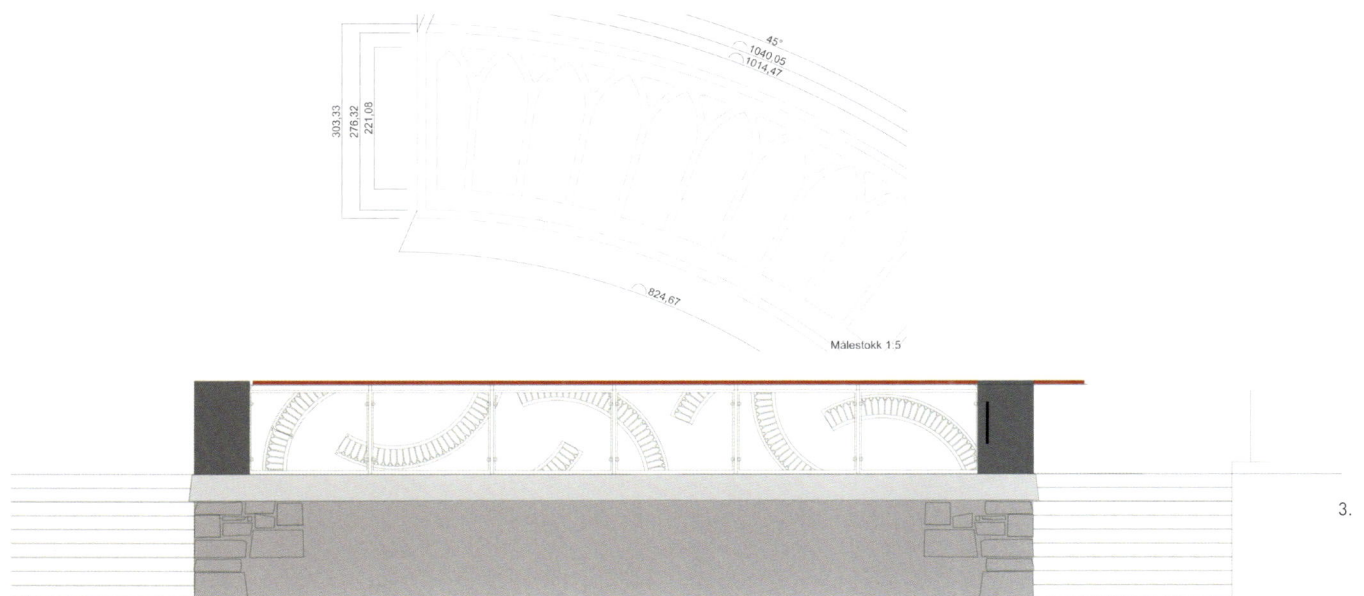

1. Acryl on canvas. Inspiration for the paving pattern.

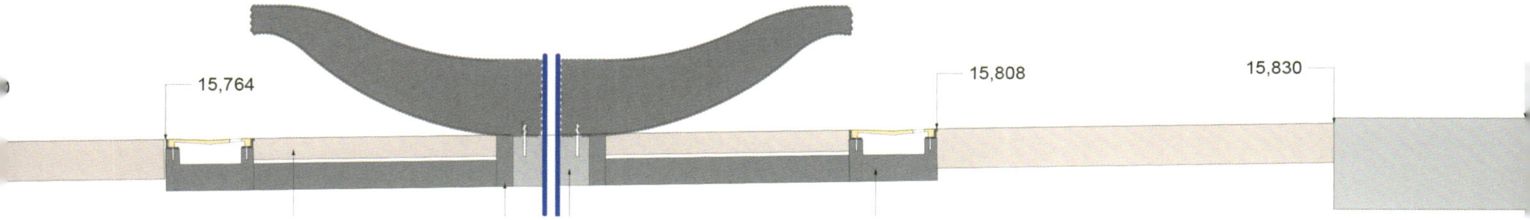

2. Construction drawing of the basin.

3. Elements in the church's facade was used as design pattern on the glass walls on the bastion.

TØNNE STREET

VEJLE, DENMARK

TØNNE STREET is a short and narrow street in the centre of Vejle. It was to become part of the main pedestrial area.

The main task was to make use of what stone resources the city had in storage. Secondly the street had to be cut off from normal traffic. Only lorries should be allowed to drive through.

The solution was to construct a large, granite sculpture in the start of the street.

Where Tønne street meets the main pedestrian area in the middle of Vejle historic centre, there was room for a small extension with just enough space for a couple of coffee tables and a maple tree.

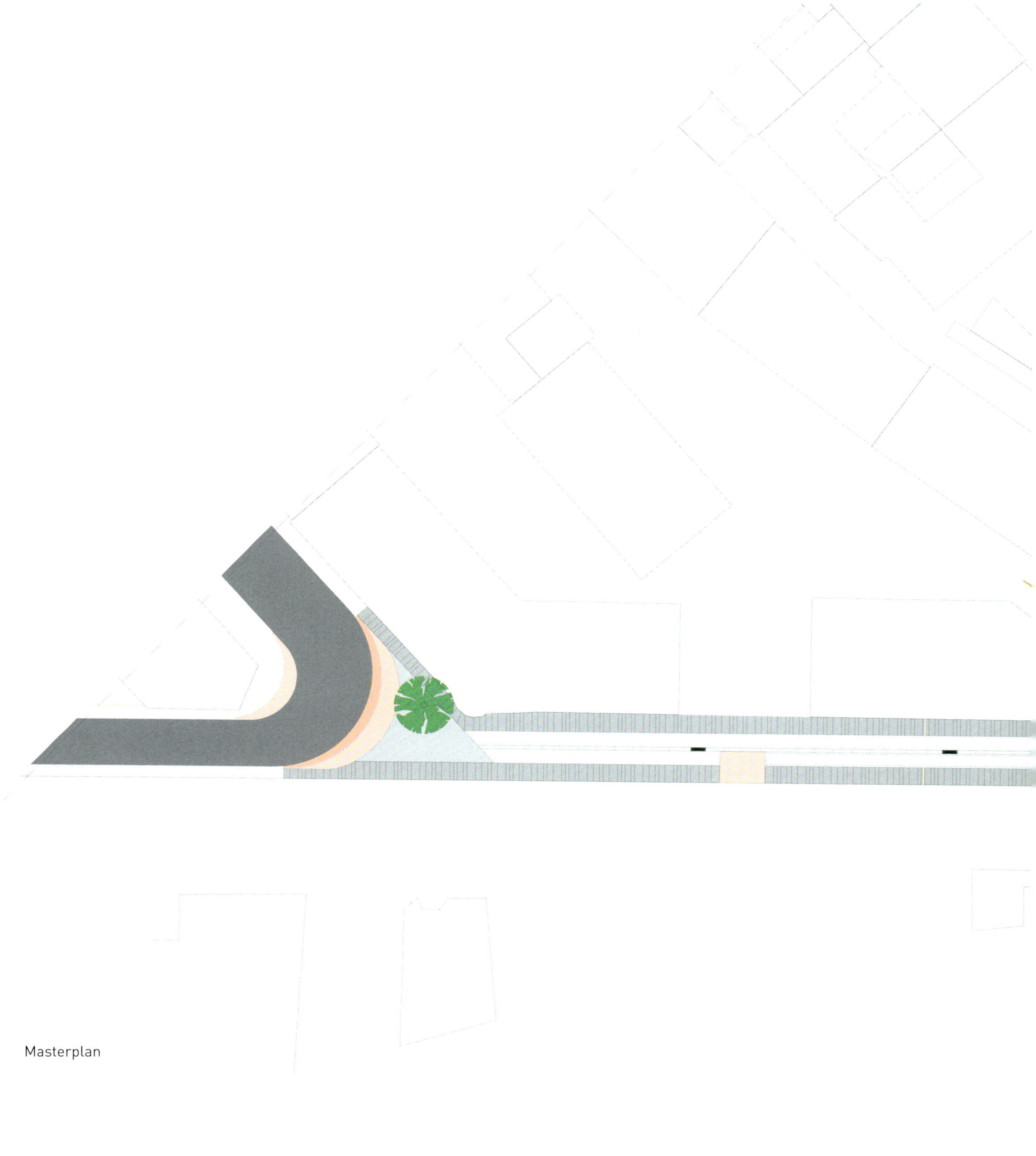

Masterplan

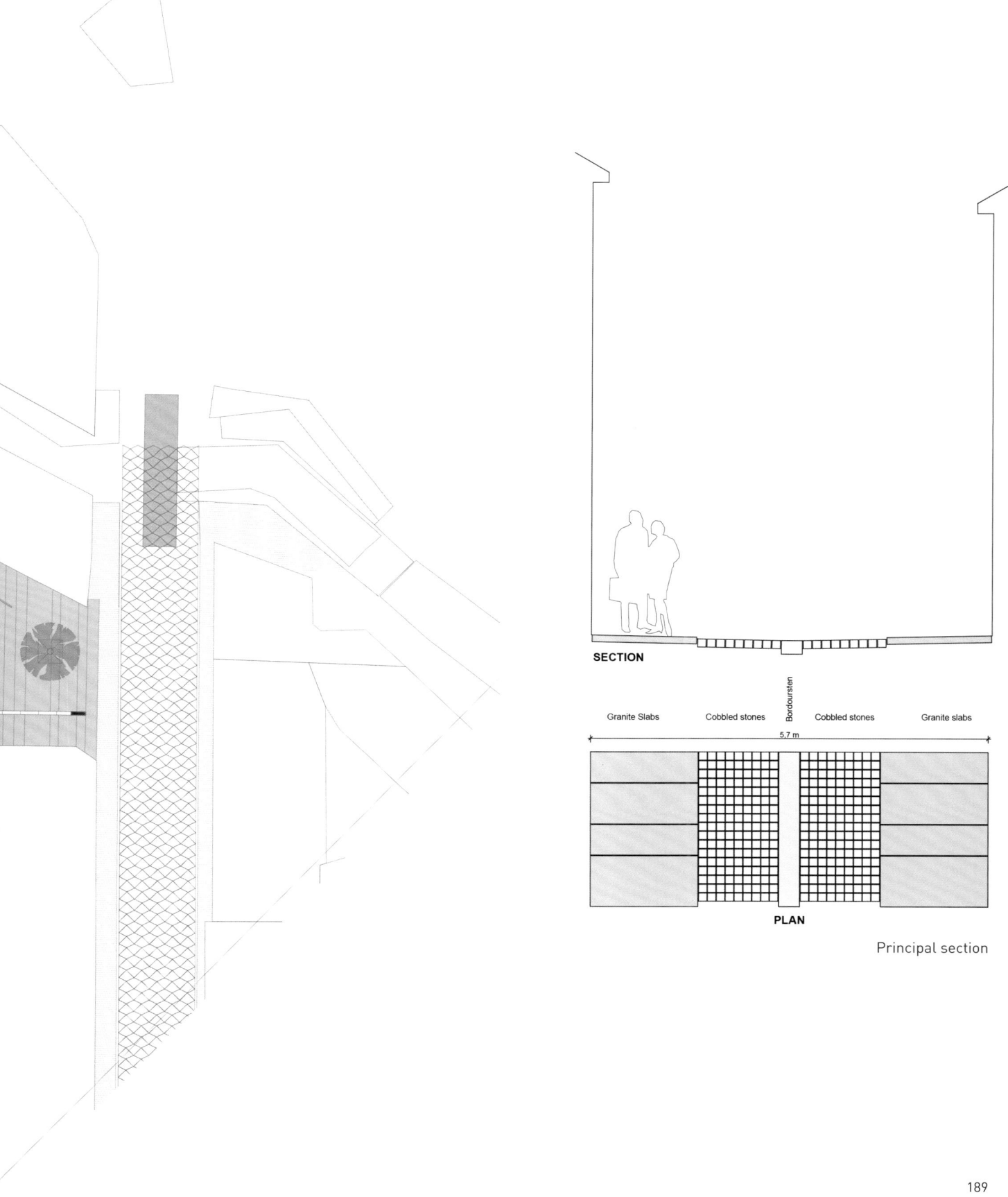

Principal section

Two alternative proposals. The one above was selected.

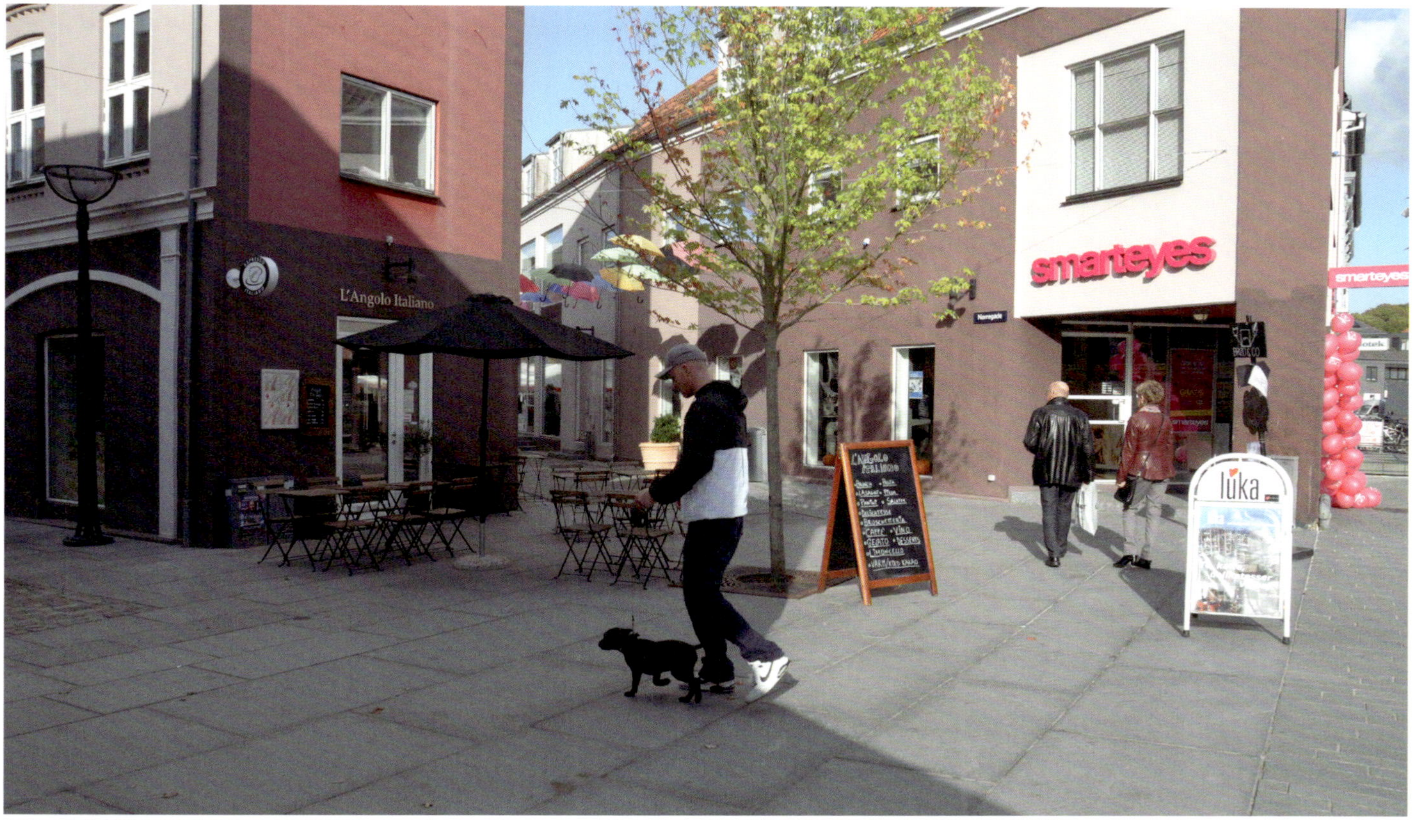
L'Angolo Italiano
smarteyes
smarteyes
luka

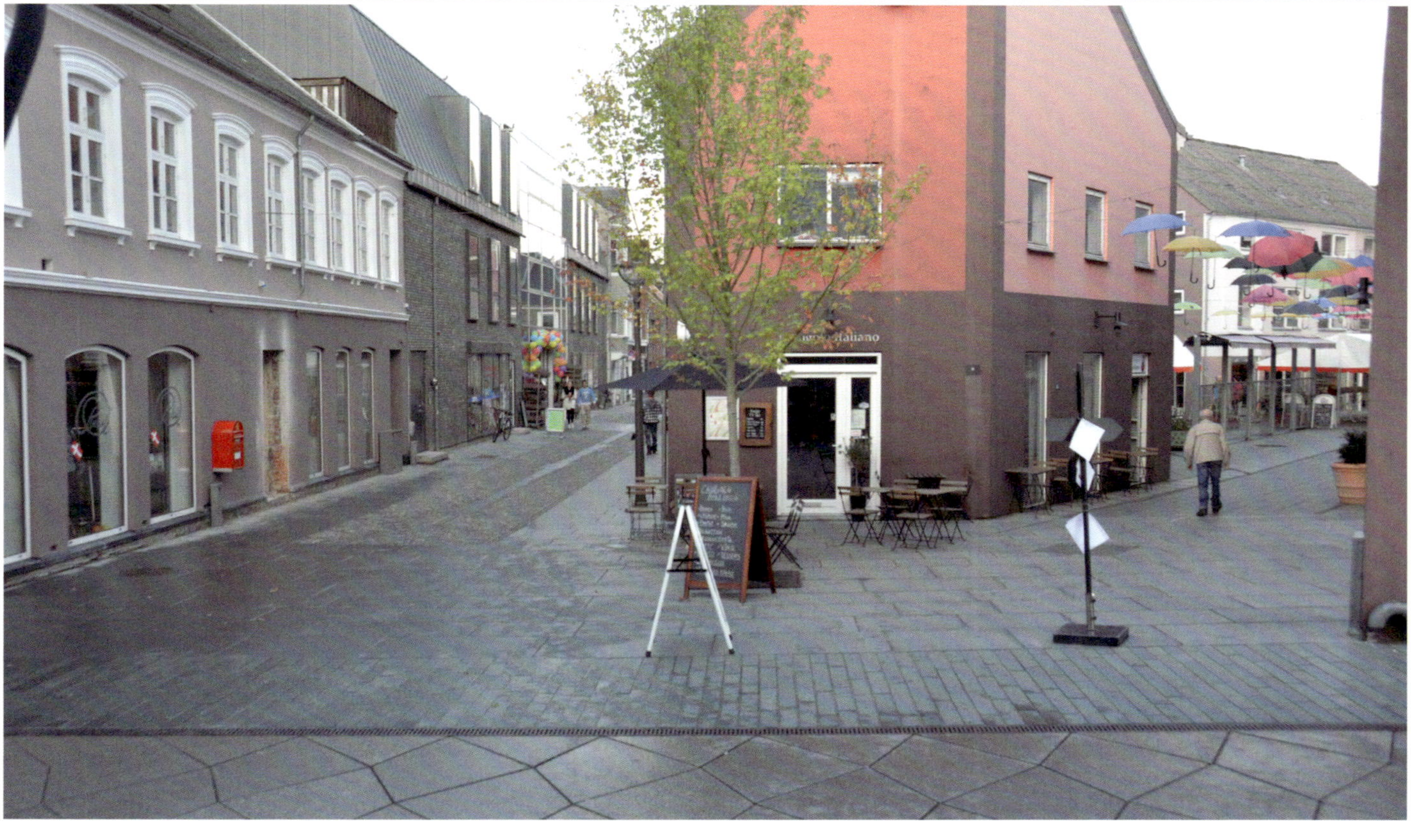

To prevent driving into the pedestrian area, a large vaulted granite form was laid out.

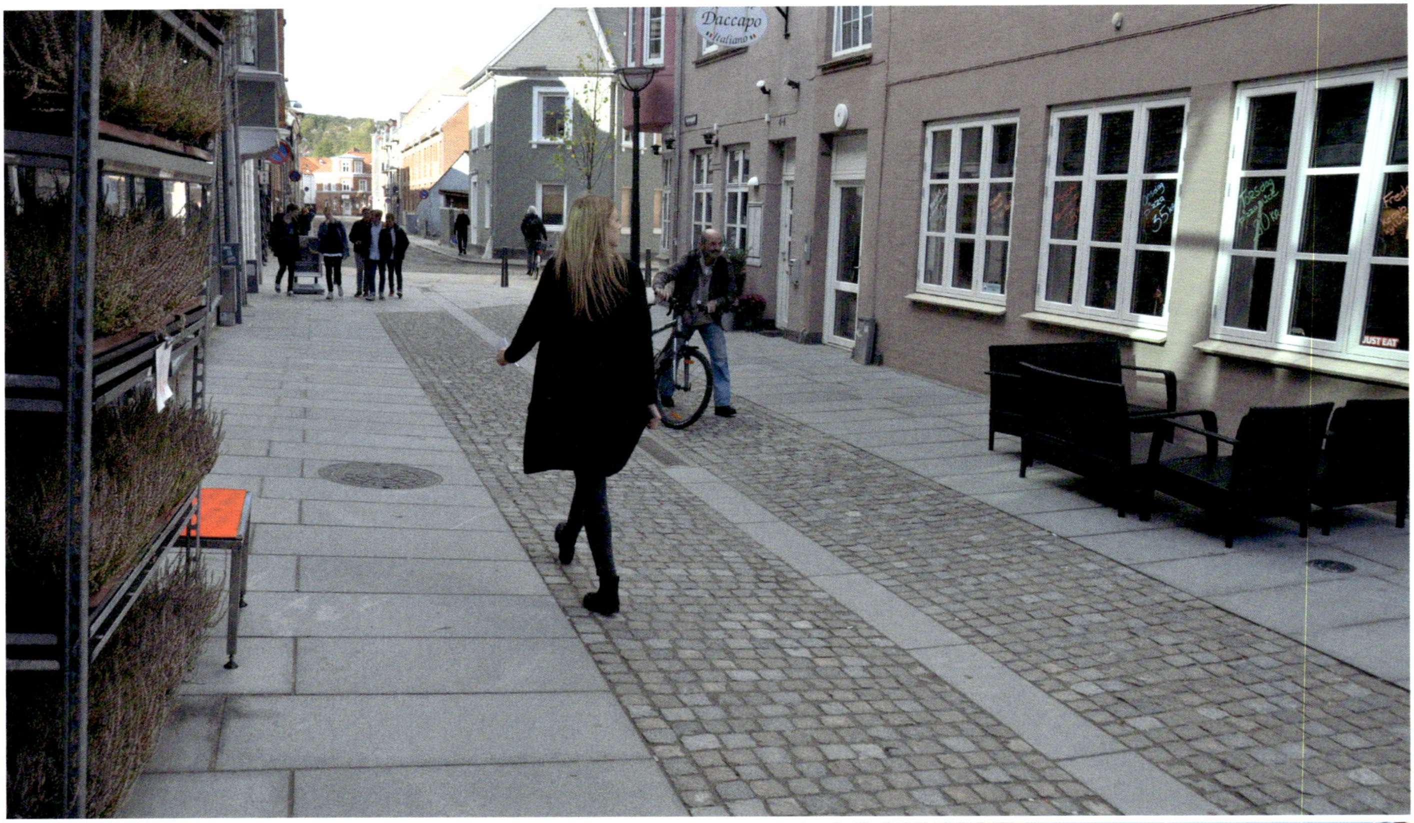
Daccapo
Italiano
JUST EAT

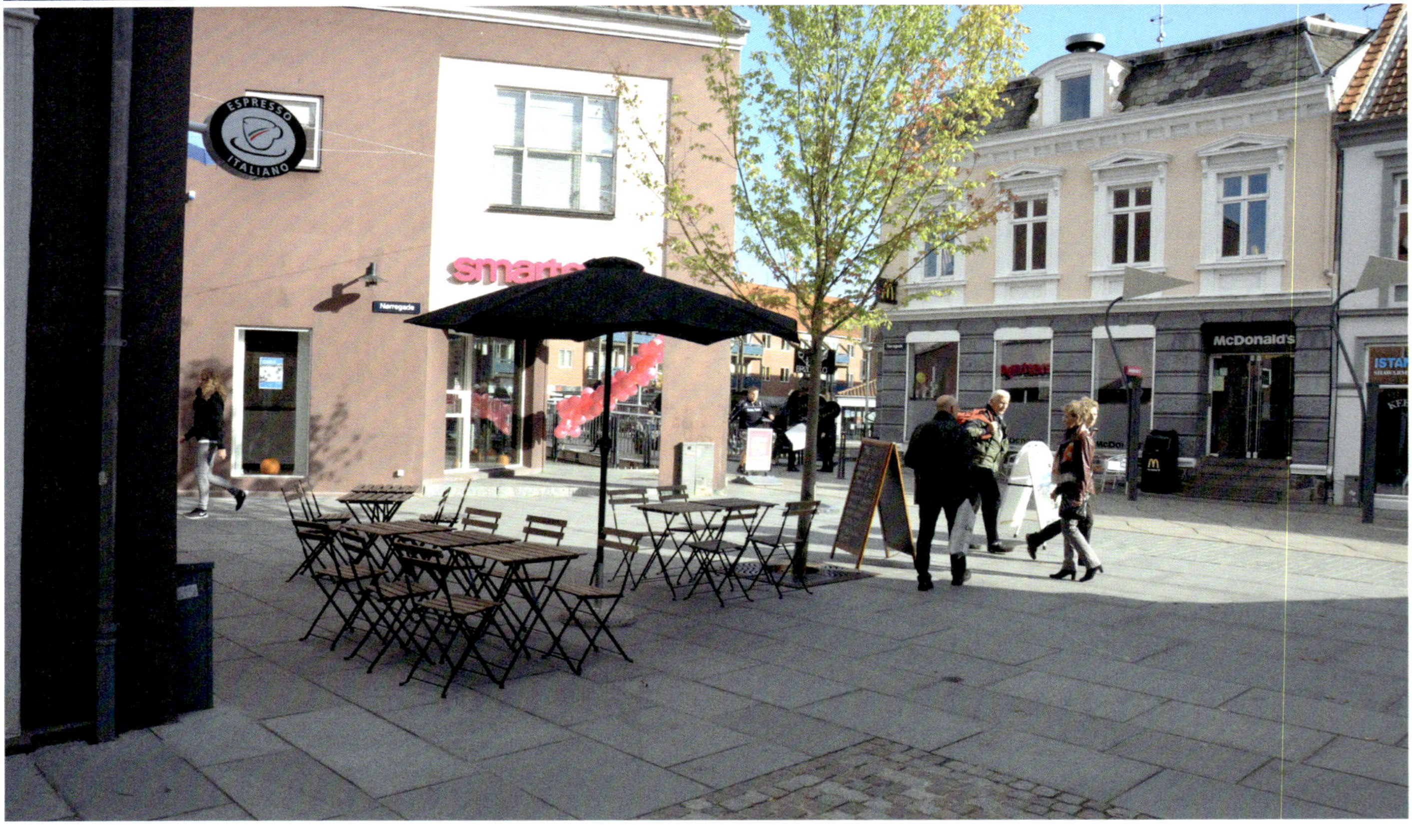
ESPRESSO
ITALIANO
McDonald's

Custom designed storm drain in bronze.

Early proposal.

VEJLE THEATRE SQUARE
VEJLE, DENMARK

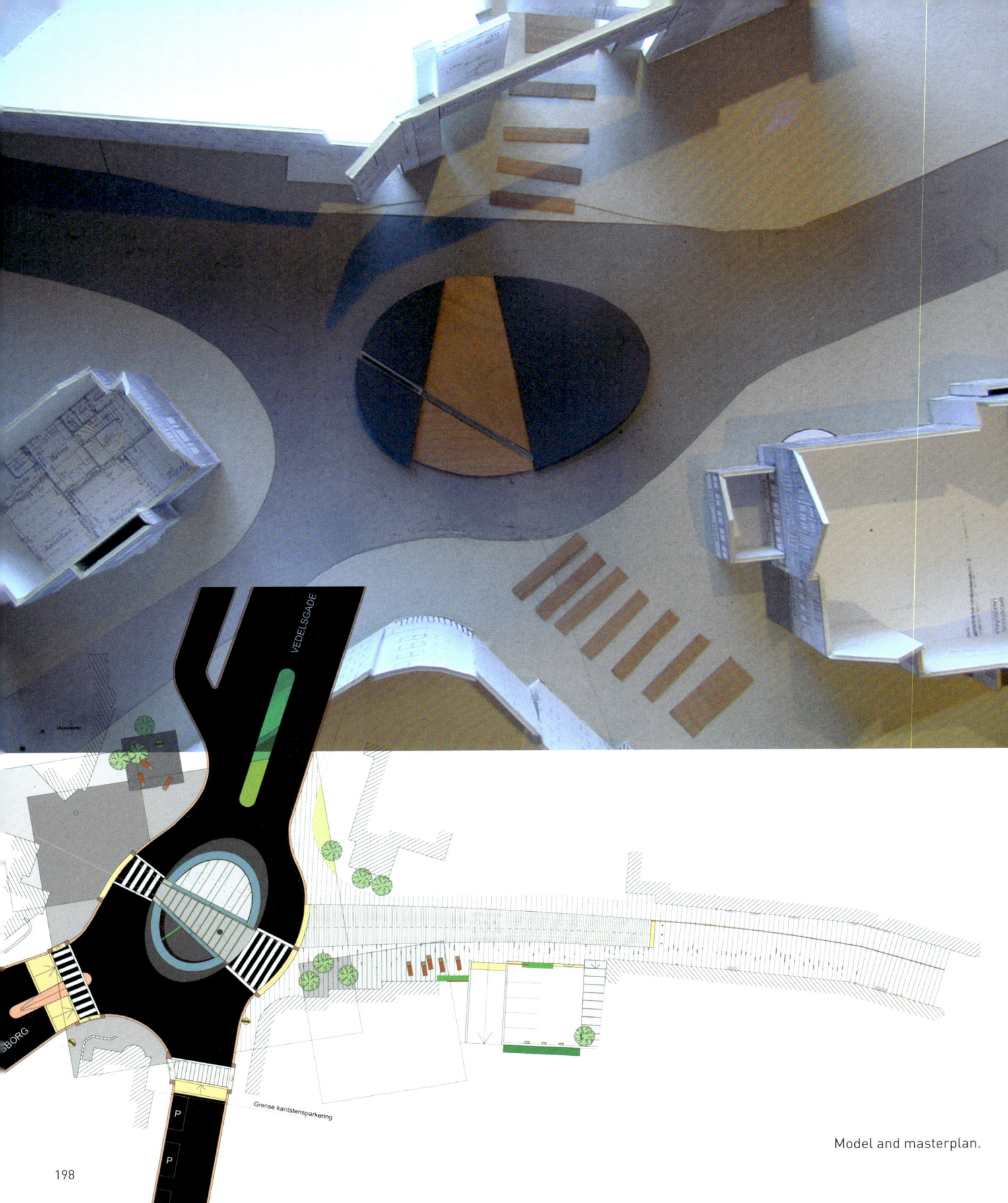

Model and masterplan.

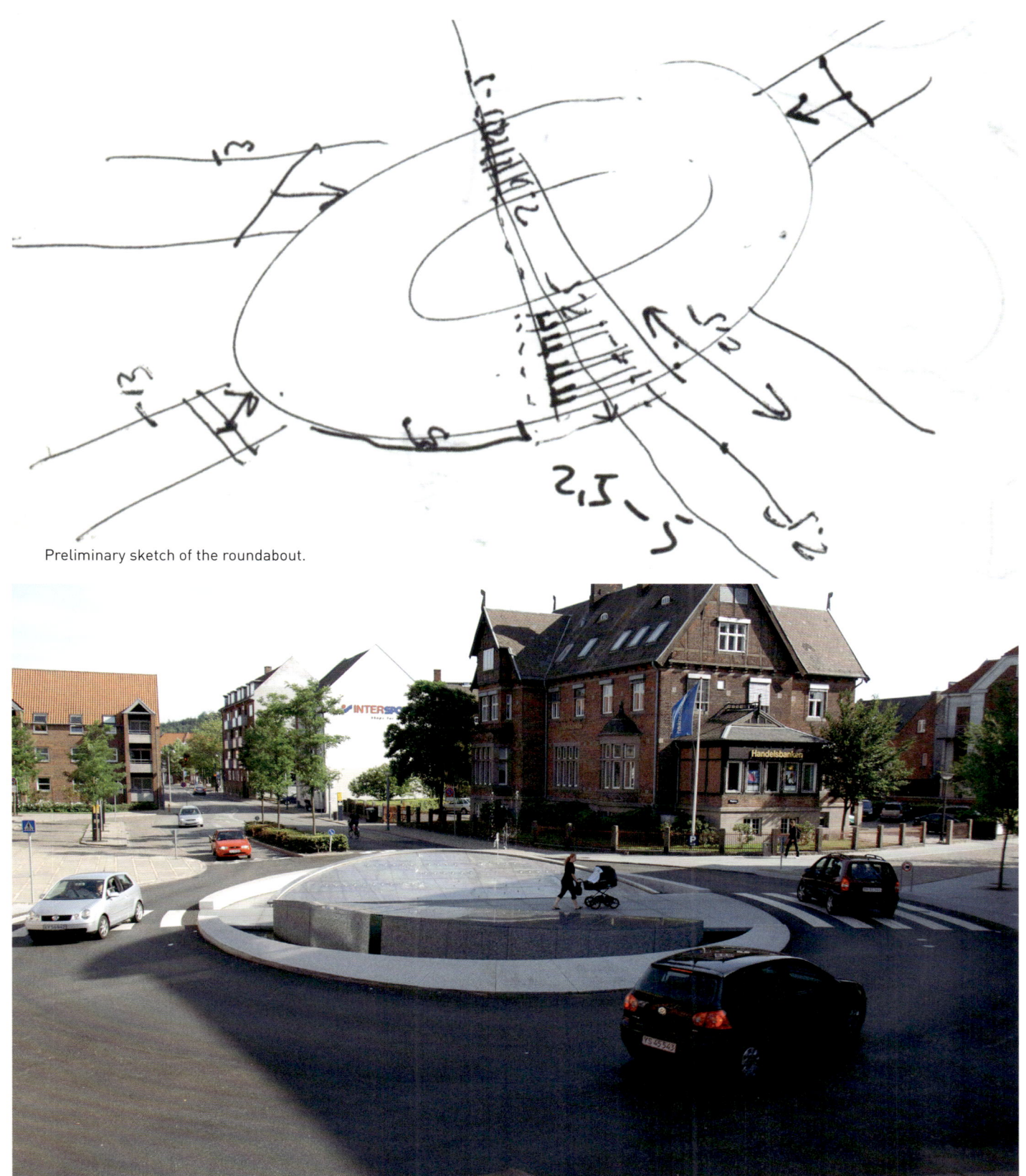

Preliminary sketch of the roundabout.

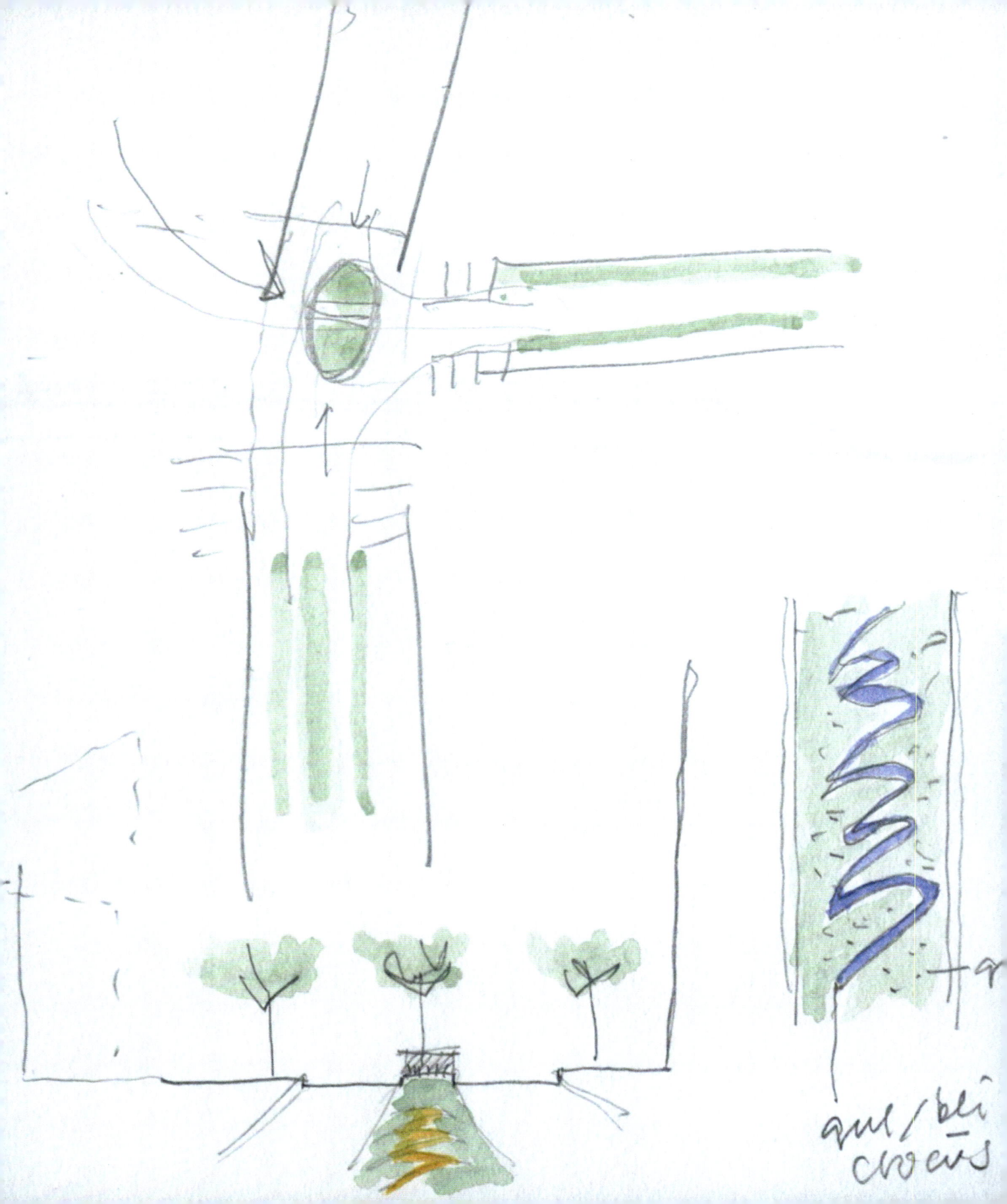

gul/blü
crocus

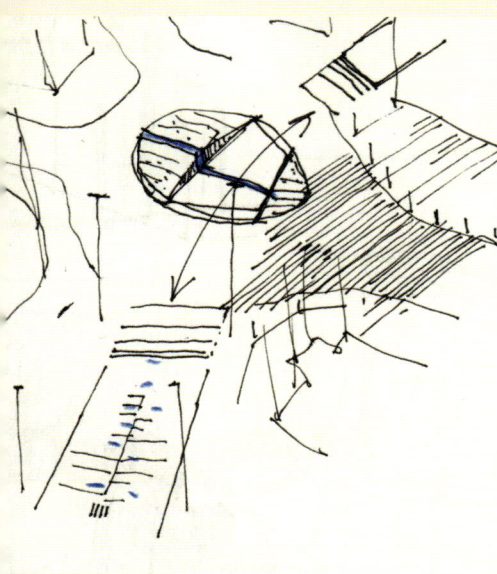

The main conceptual idea is to prepare the audience for the performance. By walking through the light and water fountains, the mind will purify and leave the city pulse behind, being ready for the art performance inside the Theatre building.

Vejle City´s main theatre is situated in the city centre. Adjacent to the theatre, lies the City Park. On the other side is a roundabout, which is transformed into a transition-zone between the city and the Theatre. The former situation was a traditional roundabout with a sunken lawn and a flower-arrangement. The cars were preferred in this situation, and the pedestrians had to play second string and should move around until a marked crossing was obtained.

Instead of these traditional pedestrian crossings, the visitors to the theatre are now invited to walk across a granite-bridge that spans the pool. A solid edging is made from light granite, to be visible in the evening as well as in the middle of the day. The edge will guard the people on this traffic island, and one really feels safe when walking across.

The square is lighted by floodlights mounted on three light poles, one large and two minor.

Inside the edge, the islands Illuminated cascades of water pour out of a steel plate that rests upon layers of glass, lighted from within. Thus the glass wall, fronting the oncoming traffic, is clearly visible in the night with its intense, white light, and gives the traffic a massive signal to turn right.

A huge granite-block, facing the other direction, is split by a massive slice of glass, which transforms the front lights from the passing cars into a carpet of light on the walkway.

The street that leads to the square, Flegborg, has limited access by car, only to a parking lot midway. To conceal the cars, two massive plates of corten steel have been raised at both ends. The plates are perforated in order to create shadow patterns on the pavement.

Sketch and model study of light emission.

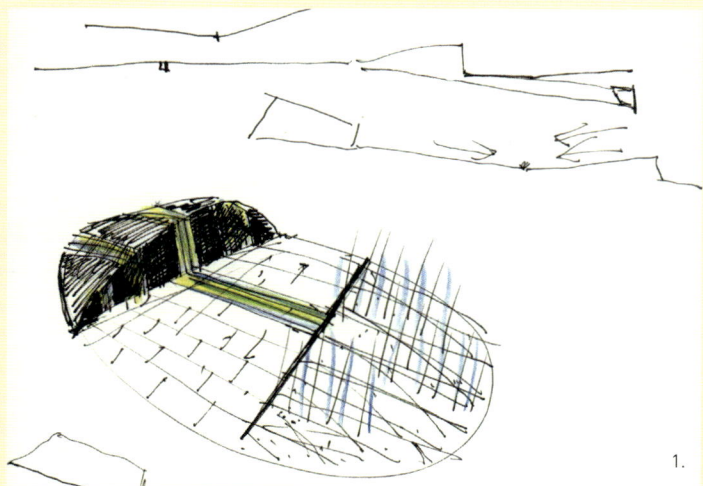

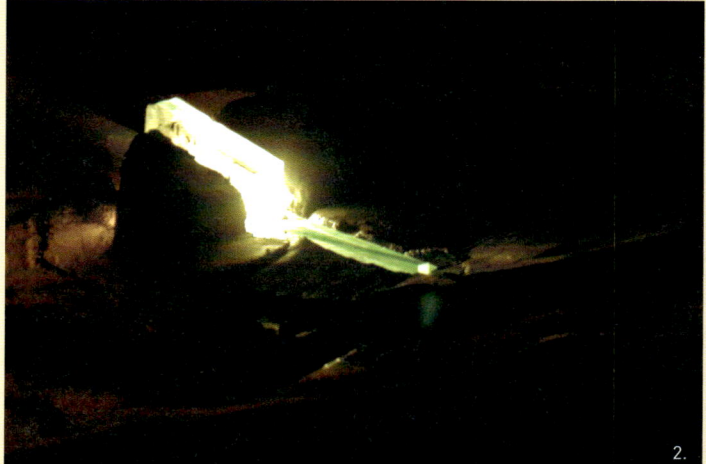

1. Sketch-idea.
2. Scale model. Light-test.
3. Production drawing of massive stone element.

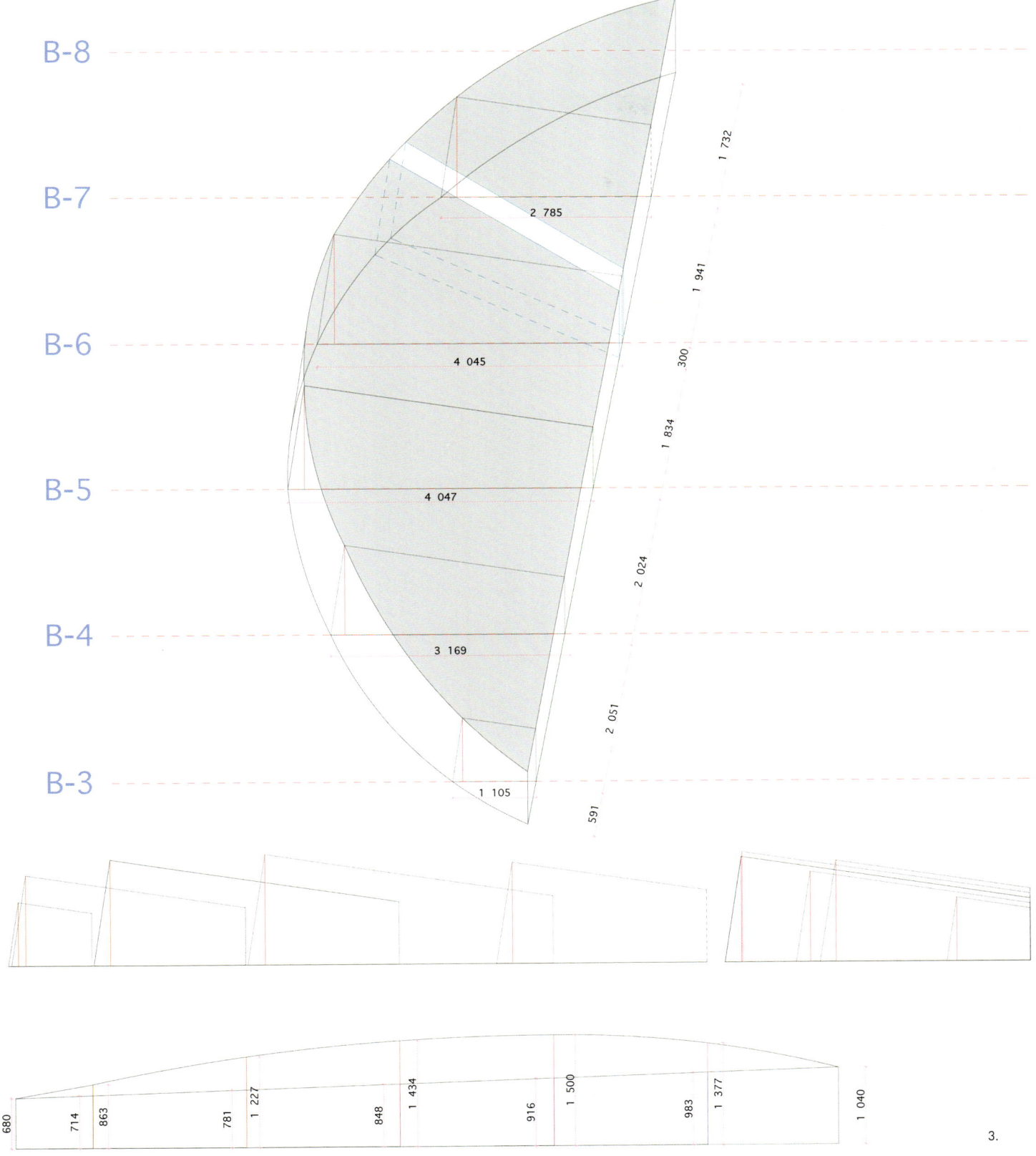

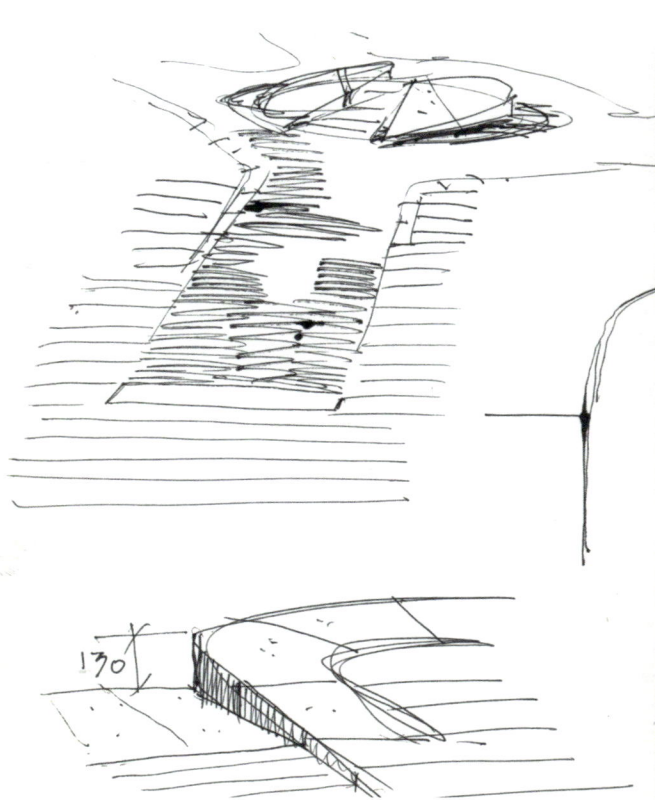

1. Massive granite element.
2. Section of fence in massive corten steel plate.
3. Sketch of fence.
4. Corten fence.

1.

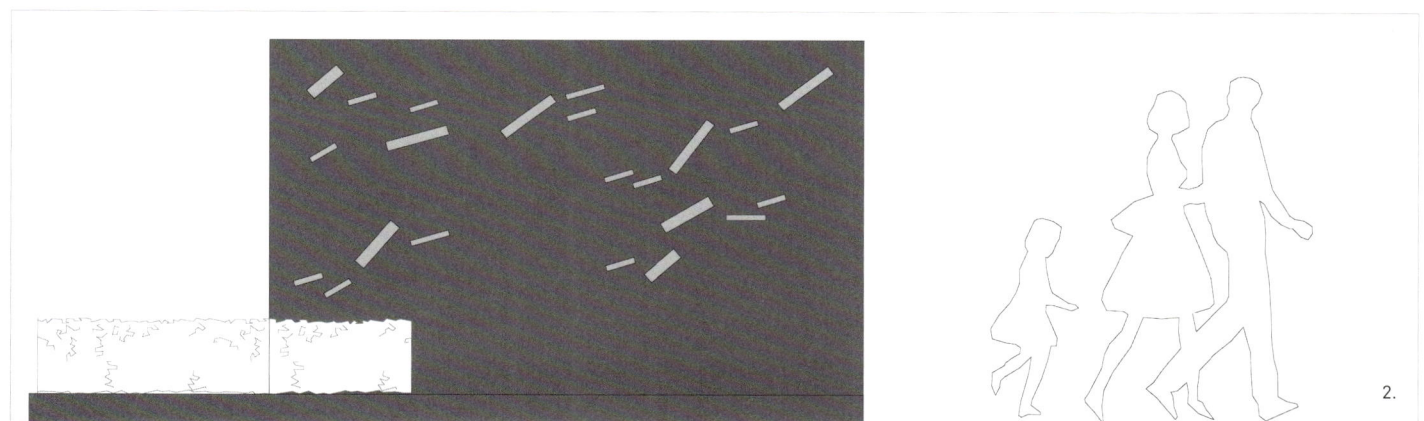

2.

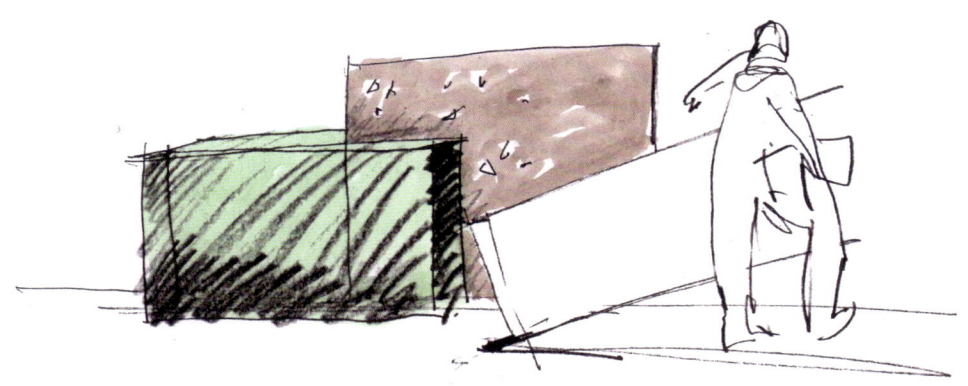

3.

4.

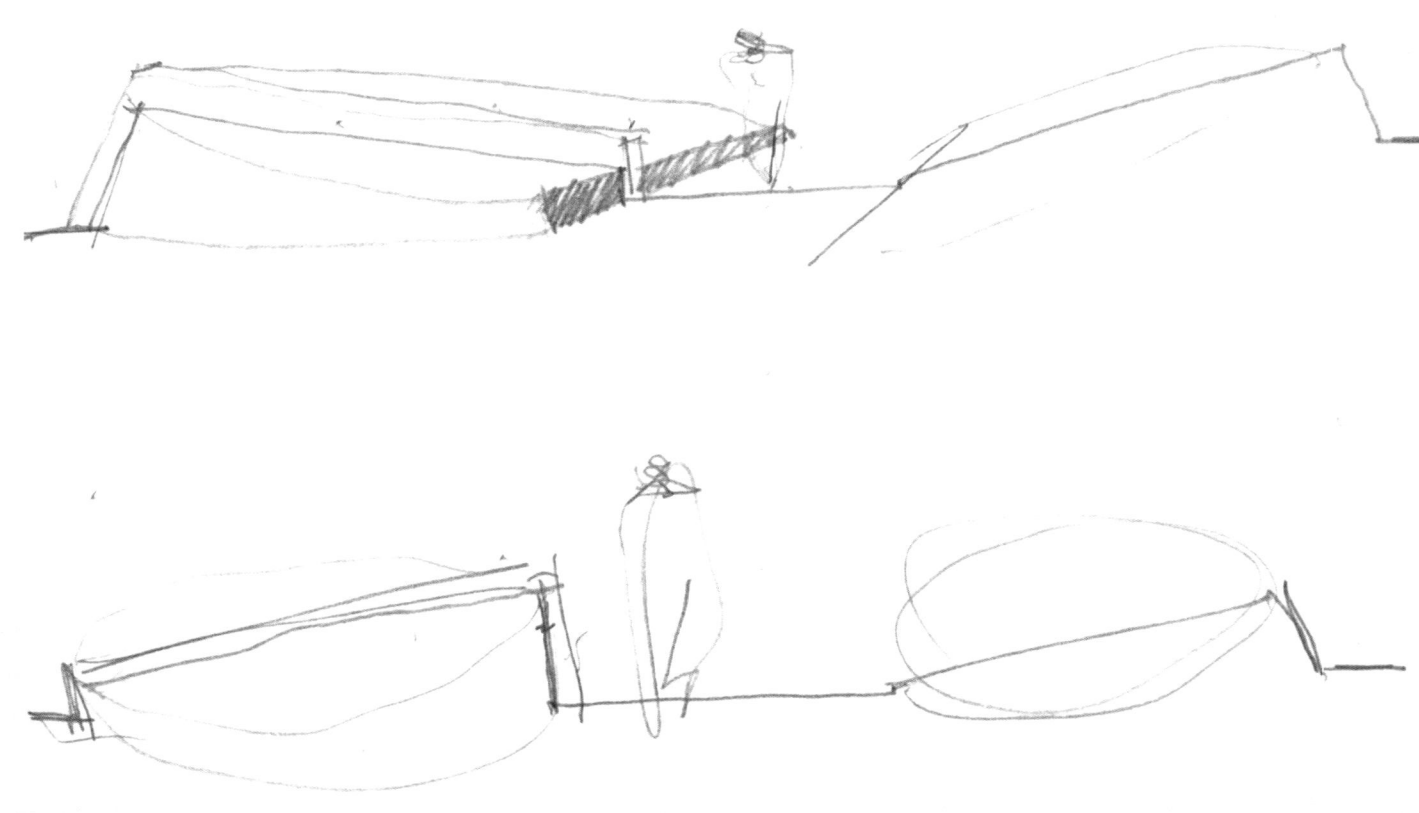

Four sketches showing principle access through the roundabout, fountain and of the maintenance-room below.

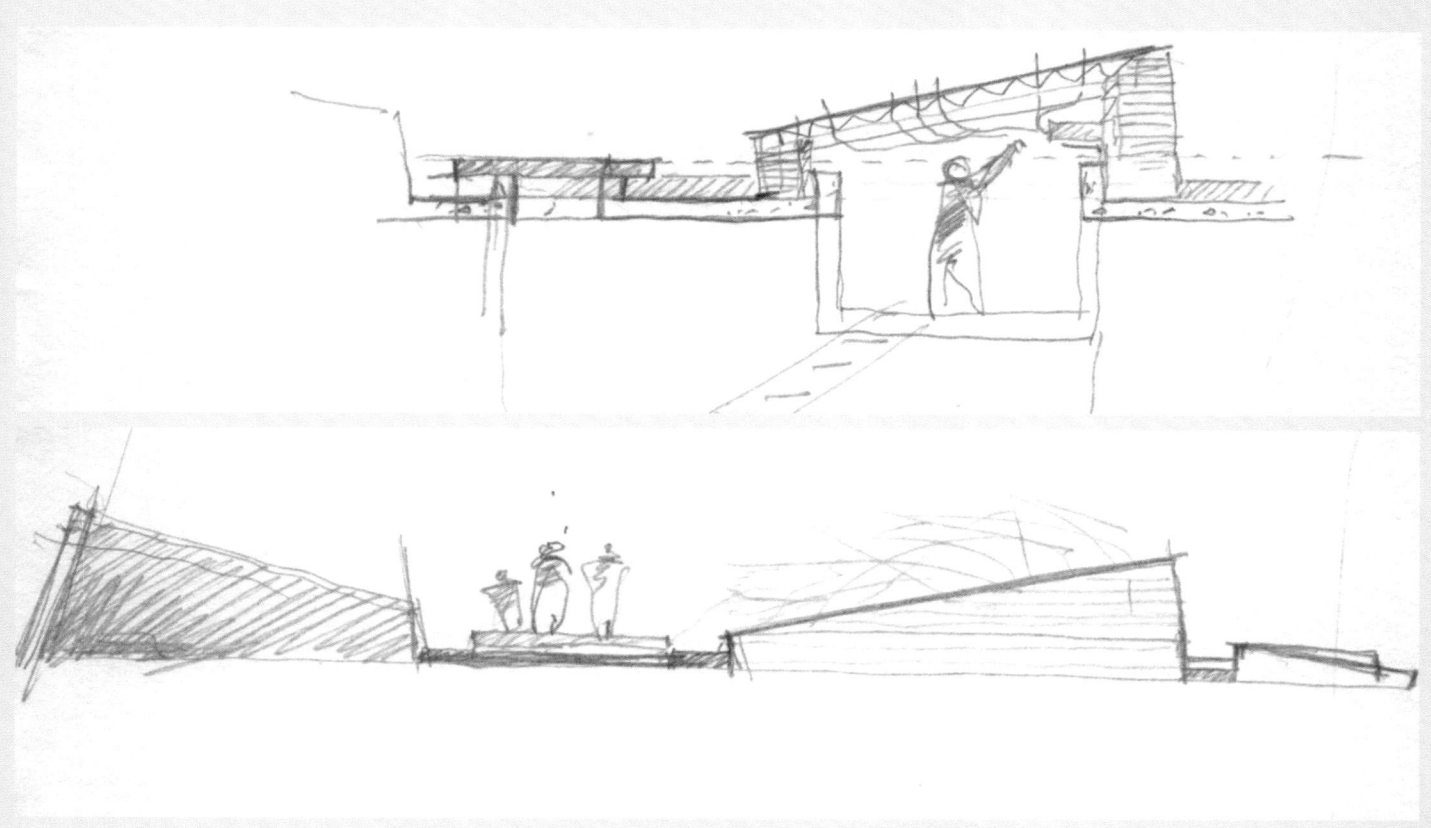

VINCENT LUNGE SQUARE
BERGEN, NORWAY

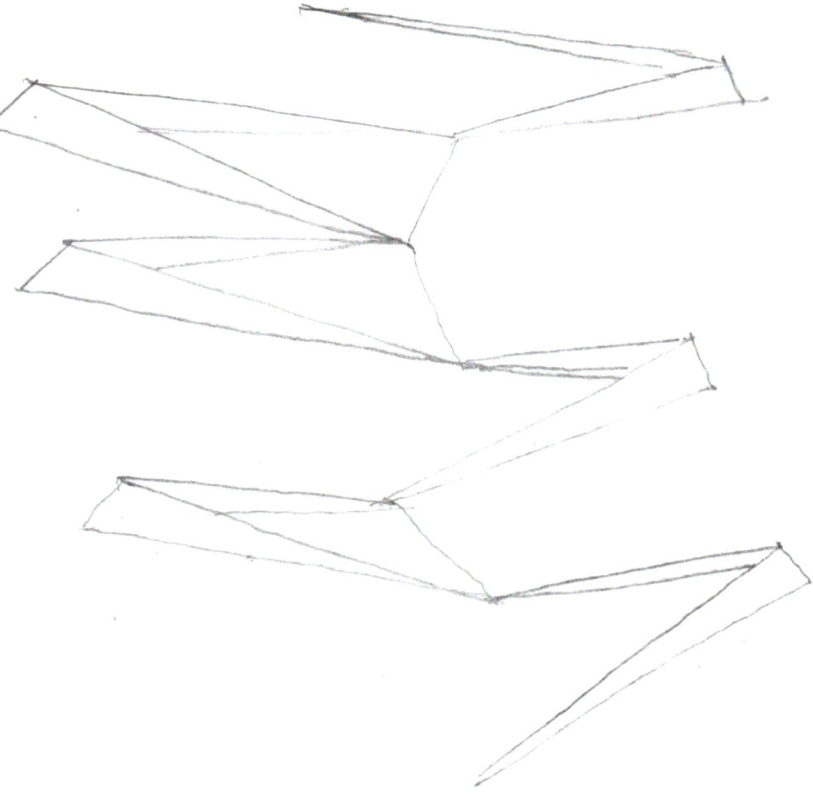

A 100 meter long zig-zag granite-form is dividing the square in two.

The outside in black basalt facing the heavy bus-traffc.

The inside in white granite is the calmer zone for pedestrians

Today thousands of pedestrians cross the square each day. When the reconstructed city centre expansion moves on farther to the south, this number will soon multiply.

The design idea is to separate the bus traffic on one side from the pedestrians on the other by a tilted zig-zag form in polished black granite. Several hundred people are looking down at the square from the adjacent buildings. Therefore it was crucial to compose the square as a large scale painting, the details will not be visual until on reach the lower sections of the buildings.

Vincent Lunge Square is a highly urban square in Bergen´s new financial district, though the site is quite old. During the Reformation-period around 1550, the site was still an old Abbey, but the area was taken by the Crown soon after.

Vincent Lunge was the king´s representative in Bergen at that time, and he used the Abbey as his residence. The square is thus named after him.

Two triangles within the granite sculpture are filled with lawns facing the afternoon sun, and a third contains a fountain on a steel frame. The fountain has 21 water jets that are lighted from below.

Over the structure a loft of London Planes is rapidly developing.

The main concept is a hundred meter long, huge black, polished granite sculpture, dividing the square in two, one black side and one white side. The black traffic side consists of basalt, the white pedestrian inner part of light grey granite.

Both material have three different surfaces: Rough picked, fine picked and flamed. All slabs are 60cm wide, but they vary in length.

The main office block to the south defines the divide; the axis in the centre of the building is prolonged to it either reach the curbstone in the south, or the zig-zag in the north.

The square is lighted from lightpoles along the curbstone, and from downlights on the building to the east. The zig-zag sculpture is resting on a LED-cable, thus it literally "floats in the night"

Three-grilles, tree-protections, bicycle-racks and benches are all custom design, made from stainless steel and oak.

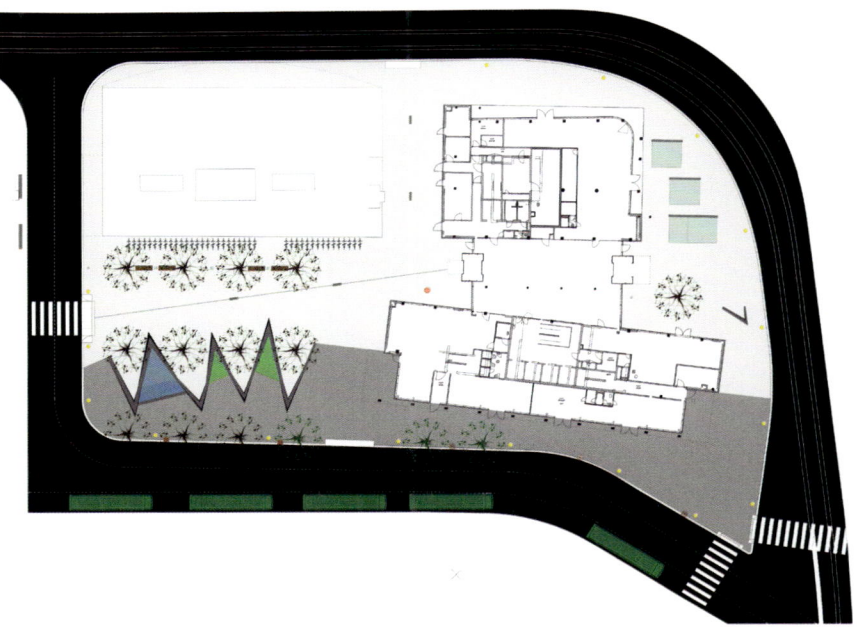

1.

220

© Arne Sælen

221

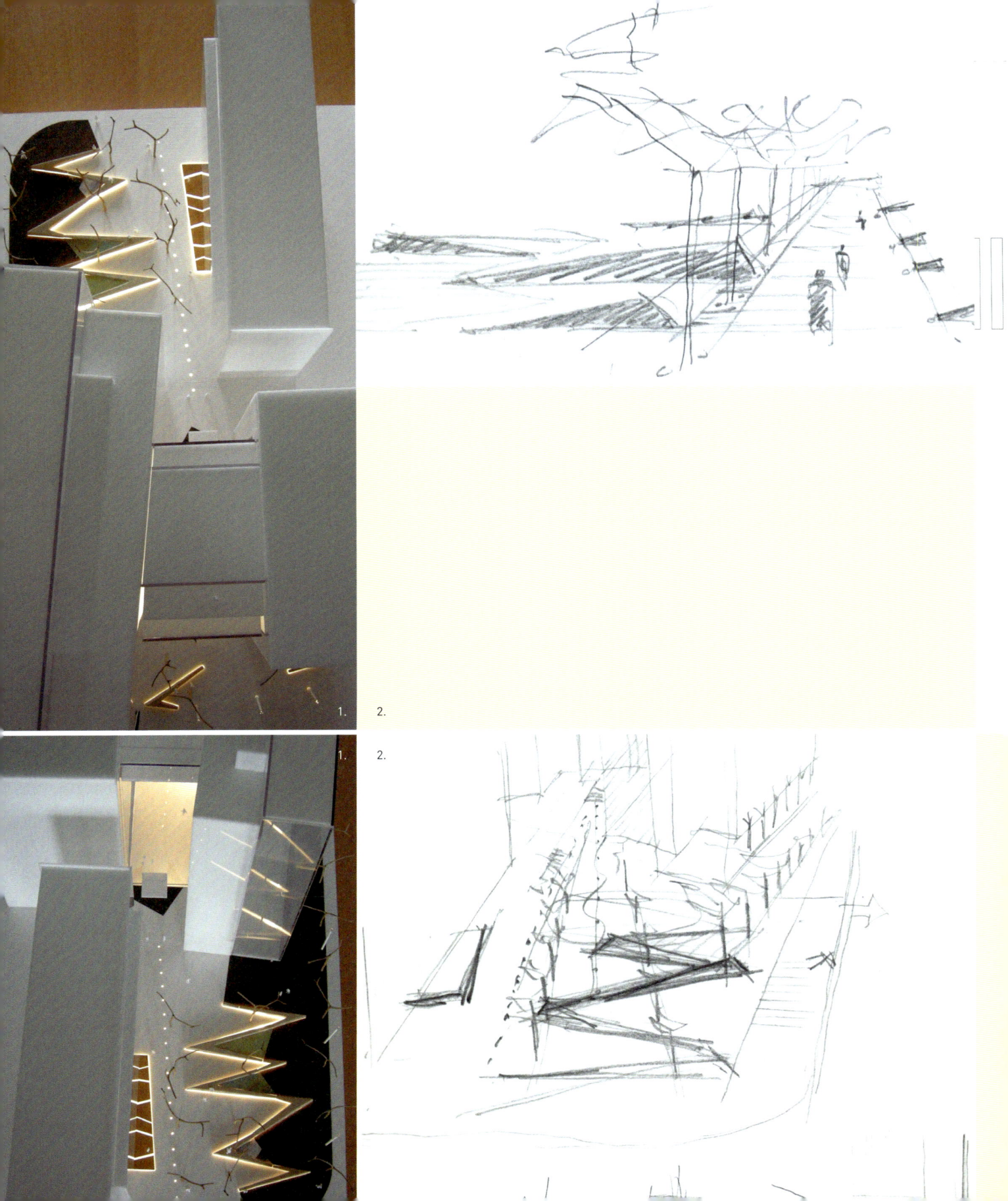

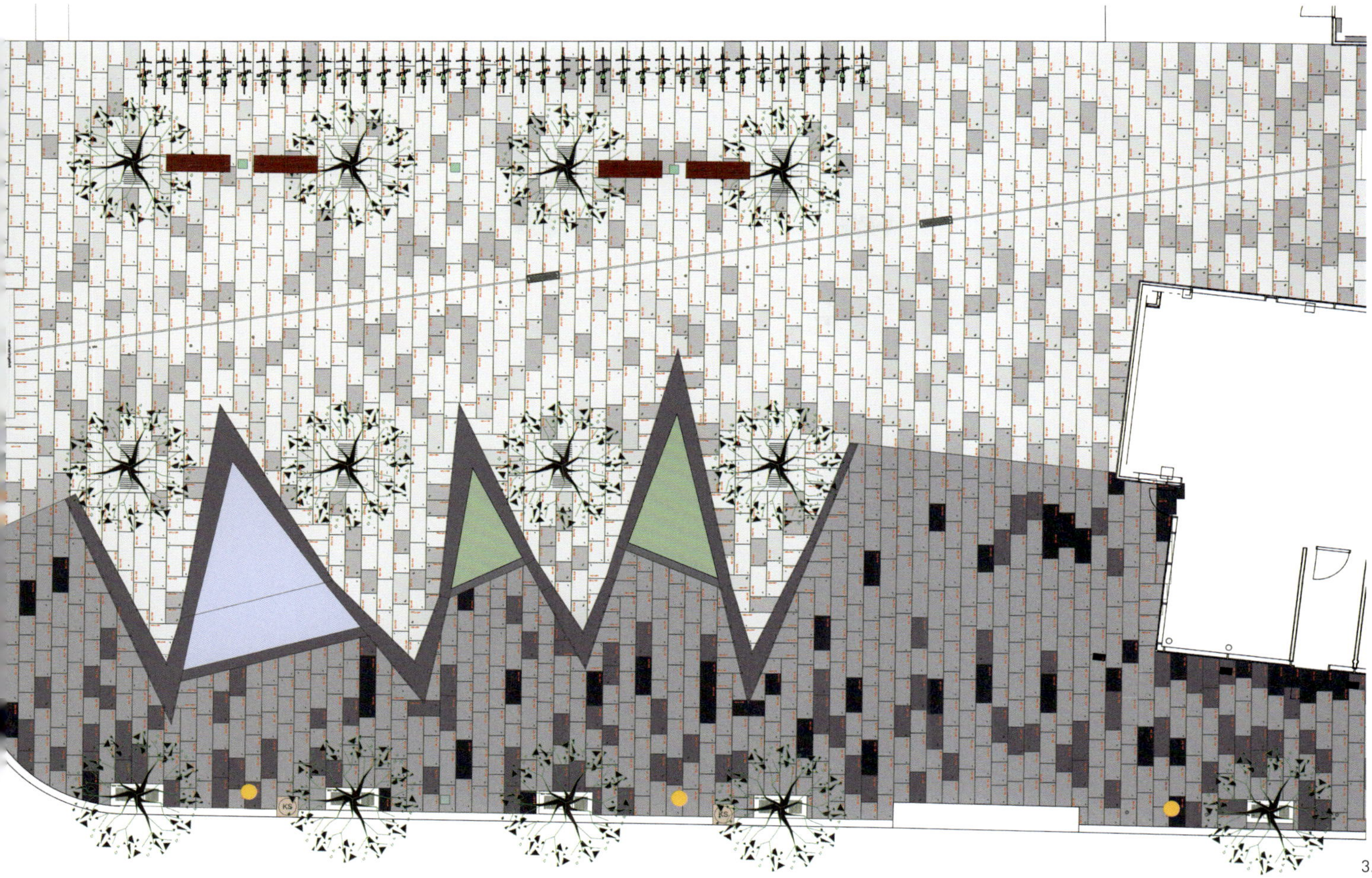

1. Scale-model 1:100 of the square with adjacent buildings.
2. Early sketches.
3. Construction plan.

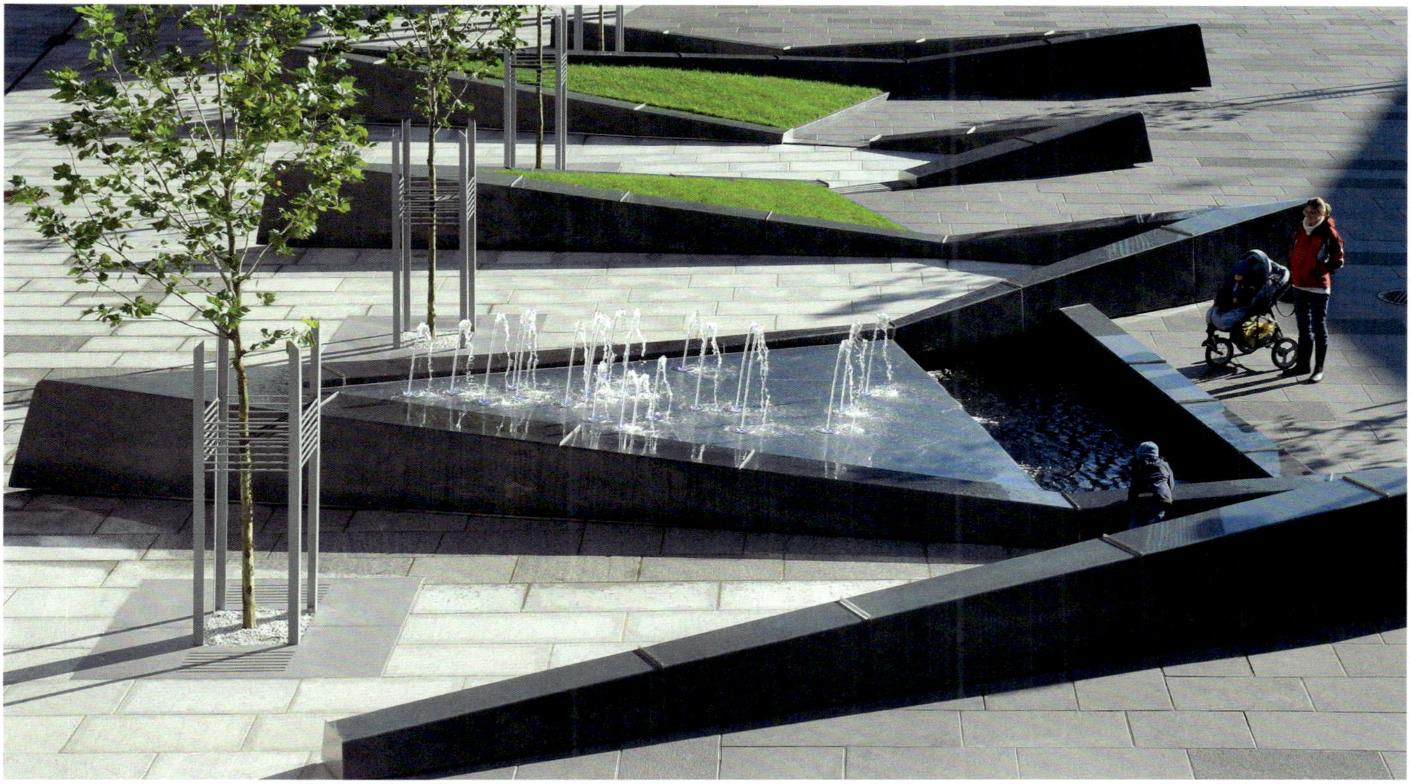

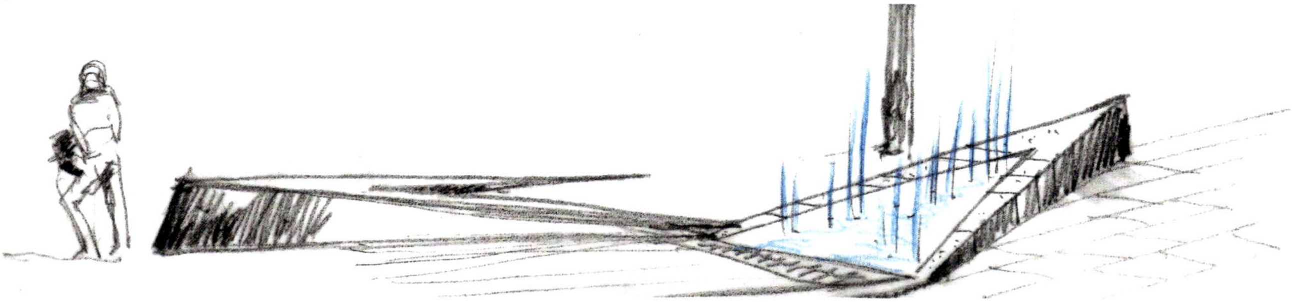

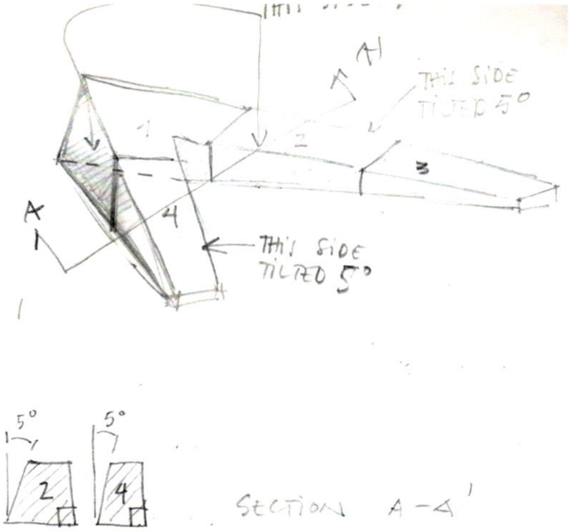
THIS SIDE
TILTED 5°
A
1
2
3
4
THIS SIDE
TILTED 5°
5°
5°
2
4
SECTION A-A'

Credit to:

All sketches and models are made by Arne Sælen
Most of the images are taken by Arne Sælen
The drone images by Ulrik Holsen

Lasse Berntzen
p.35, p.46

Christine Nundal
p.215, p.217 l, p.228 t

Hanne C Olsen
p.60 l, p.83, p.215, p.216 r, p.219 r, p.225 r, p.225 r, p.228 l

Bent René Synnevåg
p.56, p.57, p.58 l, p.65, p.102, p.174, p.180 t, p.184 t, p.216 t, p.222 r, p.223, p.224 l, p.225 t

Kari Aasen
p.47 top r, p.89, p.91 l

Special thanks to:

Asbjørn Andresen
Lasse Berntzen
Ola Bettum
Thale Bjørnerheim
Kurt Edvin Bilx Hansen
Peter Bonnén
Kristian Blystad
Henning Bødtker
Erik Dahl-Rasmussen
Trygve Eriksen
Arvid Fjæren
Gitte Frøkjær
Magne Furuholmen
Mogens Lock Hansen
Ingrid Haukeland
Odd Hylleseth

Henrik Iversen
Svein Jacobsen
Pontus Kjerrman
Ingebjørg Larsen
Jørgen Lindberg
Peer Lehn-Pedersen
Cristina R Maier
Cathrine Maske
Sofie Mellegaard
Mette Molden
Morten Molden
Arne Mølgaard
Cato Mørner
Hanne Pollen
Ulla Sandgaard
Elin Strandenes

Morten Stræde
Solfrid Stølen
Gunnar Staalesen
Christian Sunde
Ruth Svellingen
Petter Søiland
Eli Oftedal Sømme
Axel N Sømme
Gitte Fuhr Thomsen
Eli Veim
Ragnar Vaagbø
Odd-Arne Vagle
Knut Wiborg
Kristin Aarskog
Kari Aasen
Henrik Stjernholm

URBAN SKETCHES

Editorial project:
© 2023 booq publishing, S.L.
c/ Domènech, 7-9, 2º 1ª
08012 Barcelona, Spain
T: +34 93 268 80 88
www.booqpublishing.com

ISBN 978-84-9936- 643-2

Editorial coordinator:
Paco Asensio

Editor:
Arne Sælen

Layout:
Mireia Casanovas

Text by:
Arne Sælen

Printing in Spain

booq affirms that it possesses all the necessary rights for the publication of this material and has duly paid all royalties related to the authors' and photographers' rights. booq also affirms that is has violated no property rights and has respected common law, all authors' rights and other rights that could be relevant. Finally, booq affirms that this book contains neither obscene nor slanderous material.
The total or partial reproduction of this book without the authorization of the publishers violates the two rights reserved; any use must be requested in advance. In some cases it might have been impossible to locate copyright owners of the images published in this book. Please contact the publisher if you are the copyright owner in such a case.